T0236978

SpringerBriefs in Electrical and Computer Engineering

Control, Automation and Robotics

Series Editors

Tamer Başar, Coordinated Science Laboratory, University of Illinois at
Urbana-Champaign, Urbana, IL, USA
Miroslav Krstic, La Jolla, CA, USA

SpringerBriefs in Control, Automation and Robotics presents concise summaries of theoretical research and practical applications. Featuring compact, authored volumes of 50 to 125 pages, the series covers a range of research, report and instructional content. Typical topics might include:

- a timely report of state-of-the art analytical techniques;
- a bridge between new research results published in journal articles and a contextual literature review;
- a novel development in control theory or state-of-the-art development in robotics;
- an in-depth case study or application example;
- a presentation of core concepts that students must understand in order to make independent contributions; or
- a summation/expansion of material presented at a recent workshop, symposium or keynote address.

SpringerBriefs in Control, Automation and Robotics allows authors to present their ideas and readers to absorb them with minimal time investment, and are published as part of Springer's e-Book collection, with millions of users worldwide. In addition, Briefs are available for individual print and electronic purchase.
Springer Briefs in a nutshell

- 50–125 published pages, including all tables, figures, and references;
- softcover binding;
- publication within 9–12 weeks after acceptance of complete manuscript;
- copyright is retained by author;
- authored titles only—no contributed titles; and
- versions in print, eBook, and MyCopy.

Indexed by Engineering Index.

Publishing Ethics: Researchers should conduct their research from research proposal to publication in line with best practices and codes of conduct of relevant professional bodies and/or national and international regulatory bodies. For more details on individual ethics matters please see:
https://www.springer.com/gp/authors-editors/journal-author/journal-author-helpdesk/publishing-ethics/14214

More information about this subseries at http://www.springer.com/series/10198

Jerome Le Ny

Differential Privacy
for Dynamic Data

 Springer

Jerome Le Ny
Department of Electrical Engineering
Polytechnique Montréal
Montreal, QC, Canada

ISSN 2191-8112   ISSN 2191-8120   (electronic)
SpringerBriefs in Electrical and Computer Engineering
ISSN 2192-6786   ISSN 2192-6794   (electronic)
SpringerBriefs in Control, Automation and Robotics
ISBN 978-3-030-41038-4   ISBN 978-3-030-41039-1   (eBook)
https://doi.org/10.1007/978-3-030-41039-1

MATLAB is a registered trademark of The MathWorks, Inc. See mathworks.com/trademarks for a list of additional trademarks.

Mathematics Subject Classification (2010): 93E11, 93E10, 93A17, 68P30, 94A60

This Springer imprint is published by the registered company Springer Nature Switzerland AG
The registered company address is: Gewerbestrasse 11, 6330 Cham, Switzerland

Preface

As signal processing and control systems extend their reach to new large-scale systems that enable a more intelligent infrastructure, powering smart buildings, smart grids, or intelligent transportation systems, algorithm design in these fields faces new types of constraints due to the fact that the raw data being processed originates from people. Indeed, while individuals can benefit from more efficient energy or transportation systems that adapt to their consumption or mobility patterns, they also expect and should obtain reasonable guarantees that these systems cannot be used to infringe on their privacy. Guaranteeing privacy in cyber-physical-social systems is a fundamental constraint tied to ethical considerations (fundamental right to privacy). In addition, if we accept that one can define "levels" of privacy, then if two systems provide the same service, the one offering the higher level of privacy should have an advantage in that it will be more acceptable to human participants and therefore should see an increased level of adoption.

Research on privacy-preserving data analysis has been carried out for several decades in statistical science and more recently in computer science. It is a new research direction in systems theory, but a large and growing number of workshops, special sessions at conferences, and special issues in academic journals in the last few years confirms the growing importance of this subject for research communities that are in particular interested in dynamic systems and processing signals over time rather than analyzing static databases.

This monograph considers one important approach to the privacy-preserving analysis of dynamic data streams, which aims to enforce a formal quantitative notion of privacy called *differential privacy* on published signals. Differential privacy is a strong notion of privacy, which has an operational advantage compared to other approaches and privacy definitions in that it does not require models of privacy attacks to define the guarantee it provides. As a result, it has seen increasing adoption, including by major information technology companies and government offices publishing official statistics. It is therefore a good candidate for integration into sensor and control networks producing real-time statistics about a population of users.

The intended audience for this book ranges from students and researchers in control theory and signal processing to engineers and data scientists working with sensitive dynamic data streams and wishing to develop and implement algorithms providing rigorous privacy guarantees and still good performance to analyse and process these signals. Chapter 1 provides a general introduction to privacy-preserving data analysis and defines the notion of differential privacy, which is used in the rest of the monograph as the formal privacy guarantee to enforce. Chapter 2 then describes the first basic mechanisms that one can use to make a signal differentially private by perturbing it with an appropriate amount of additive noise. These mechanisms represent building blocks to develop more complex algorithms with better performance, starting in Chap. 3. A two-stage architecture is proposed for privacy-preserving filtering, by which a sensitive signal is pre-filtered before being perturbed by white noise, while a final filter attempts to recover the desired output signal. This architecture reappears on several occasions in the following chapters, where a common theme is how to leverage (publicly available) signal models to further improve the performance. We consider the case of stationary signal with known second order statistics in Chap. 4, signals explained by linear state-space models in Chap. 5 and finally the design of differentially private nonlinear model-based observers in Chap. 6.

I wish to thank the people who shaped my view and understanding of the topic of privacy-preserving data analysis, either through direct discussions or through their writing. In particular, I was introduced to this research area at the University of Pennsylvania during lectures given by Aaron Roth, and later developed some initial ideas in collaboration with George Pappas. Several students worked with me on this topic at Polytechnique Montreal, including Gérard Degue, Meisam Mohammady and Hubert André. Finally, I would like to thank Sandra Hirche for hosting me at her Chair for Information-Oriented Control at the Technical University of Munich during a sabbatical period that allowed me to finish writing this book, and acknowledge the support of NSERC, Polytechnique Montreal and the Alexander von Humboldt Foundation.

Montreal, Canada   Jerome Le Ny
December 2019

Contents

Notation and Abbreviations

iid	Independent and identically distributed
pdf	Probability density function
iff	If and only if
$\mathbb{N}$	Non-negative integers $\{0, 1, 2, \ldots\}$
$\mathscr{C}^1$	Continuously differentiable functions
I_d, I	$d \times d$ identity matrix, or identity matrix with the dimensions unspecified
$\mathrm{diag}(v)$	Diagonal matrix with the components of the vector v on the diagonal
$|v|_p$	p-norm for a vector $v \in \mathbb{R}^d$
S^d	Space of vector-valued discrete-time sequences $\{x_t\}_{t \geqslant 0}$ with $x_t \in \mathbb{R}^d$
$\|x\|_p$	ℓ_p-norm for a signal $x \in \mathsf{S}^d$
ℓ_p^d	Subspace of sequences $x \in \mathsf{S}^d$ such that $\|x\|_p < \infty$
Δq	Sensitivity of a query $q : \mathsf{U} \to V$ (for a given adjacency relation on U and norm on V)
$\mathrm{Lap}(b)$	Laplace distribution with mean 0 and variance $2b^2$
$\mathscr{N}(\mu, \Sigma)$	Multivariate Gaussian distribution with mean vector μ and covariance matrix Σ
$\|G\|_\infty$	H_∞-norm for a finite-dimensional linear time-invariant system G
$\|G\|_2$	H_2-norm for a finite-dimensional linear time-invariant system G
LMI	Linear matrix inequality
LMMSE	Linear minimum mean-squared error
LTI	Linear time-invariant
MIMO	Multiple input multiple output
MMSE	Minimum mean-squared error
MSE	Mean-squared error
PSD	Power spectral density
SDP	Semidefinite program
SIMO	Single input multiple output
SISO	Single input single output
WSS	Wide-sense stationary

Chapter 1
Defining Privacy-Preserving Data Analysis

1.1 Motivation

With the emergence of new paradigms such as the Internet of Things (IoT) or smart cities and the collection of massive datasets by companies about their customers, statistical analysis and signal processing are being increasingly applied to private data obtained from individuals. Indeed, one can say that the possibility to analyze vast amounts of personal data capturing information about the activities of private citizens is a foundational principle behind many of these current technology-driven trends. In many ways, these systems are expected to be very beneficial. For example, information collected about someone's shopping history can help websites make useful recommendations to other customers with similar interests, transit and energy consumption data should allow us to better manage our resources and protect our environment, and increased availability of health-related data collected through smartphones could lead to quicker disease outbreak detection and to medical breakthroughs. But at the same time, as the public becomes more familiar with the power of statistical inference, concerns are being voiced about the acquisition and use of personal data by companies and governments, e.g., due to potential price and service discrimination, and more generally about the impact of big data analysis on individual privacy (President's Council of Advisors on Science and Technology 2016; Electronic Privacy Information Center 2019; Markey 2015; McDaniel and McLaughlin 2009). In fact, the data collection practices envisioned to operate some large-scale monitoring and control systems often go against basic privacy rights (Warren and Brandeis 1890), a situation that could lead to people rightly rejecting these technologies despite their suggested benefits.

To complement and inform the proper regulation of the collection and use of datasets containing private information by public and private entities, it is desirable to rely on a mathematically rigorous theory to define what we mean by privacy-preserving data analysis and allow individuals to appropriately trade off the privacy

J. Le Ny, *Differential Privacy for Dynamic Data*,
SpringerBriefs in Control, Automation and Robotics,
https://doi.org/10.1007/978-3-030-41039-1_1

risks they increasingly incur with the benefits they can expect in return. Indeed, the fact that it is difficult to grasp intuitively the potentially far-reaching consequences of seemingly benign data collection practices makes the process of developing regulations in the absence of a formal framework all the more difficult. On the other hand, having precise definitions and quantitative measures of privacy allows us to determine the level of privacy loss incurred when our personal information is released or used for various purposes, and in any case forces a system designer to think rigorously about data privacy issues. Confidence and trust that we remain in control of our data is vital to the adoption of a number of systems at the core of intelligent infrastructure projects, which promise important societal benefits but cannot work without a strong participation of individuals. This trust can be improved by implementing privacy-preserving mechanisms within the technology involved in the data collection and analysis process, in addition to enacting appropriate regulations to enforce strict rules governing the collection, access and use personal information.

One important limitation of a purely legislative approach to protecting privacy is that information technology tends to move at a faster pace than regulations, and law enforcement cannot catch all cases of abuse once unsafe systems are deployed. Hence, in 1994, a Consumer Report survey (Who's reading your medical records 1994) found that 40% of insurers disclosed personal health information to lenders, employers, or marketers without customer permission. More recently, the Federal Trade Commission (FTC) released in 2014 a study of twelve health and fitness apps (Federal Trade Commission 2014), which found that these apps shared user data with seventy-six different third parties, including advertisers, and other recent studies such as Ackerman (2013) have similarly concluded that these apps pose significant privacy risks. It is therefore desirable to implement mechanisms protecting privacy as much as possible directly within the data processing platforms. Moreover, information technology should inform public policies in order to evaluate the existing and future risks associated to data mining together with the systematic risk mitigation techniques available.

1.2 Privacy Attacks

Naive attempts to protect individual privacy, for example by simply removing obvious means of identification such as names in the records of a dataset, typically fail to properly evaluate the risks of data disclosure. This is because they generally do not anticipate or underestimate the power of *linkage attacks*, by which a published dataset is cross-correlated with other available and possibly non-anonymous data to infer new information about specific individuals. One famous example of linkage attack is the re-identification of records contained in a published medical database by Sweeney in 1997 (Sweeney 1997). Sweeney pointed out that when the medical dataset was shared, explicit identifiers such as names, addresses or social security numbers were removed, but the remaining data was sufficiently rich to easily permit re-identification. In this specific example, the database contained the gender, date of birth, and ZIP code of the participants, information which turns out to be unique to a large proportion of the U.S. population. As a result, large-scale re-identification attacks could be performed by comparing the database records to voters' registration records, which also contain these three pieces of information together with the names of individuals. But more simply, uniqueness of records means that if one knows the birthdate and ZIP code of a neighbour or co-worker for example, one can immediately check if that person is in the database, which by itself might already have significant negative consequences: for example, the database could contain only the records of people suffering from a certain disease.

More recently, Narayanan and Shmatikov again presented re-identification attacks on large datasets (Narayanan and Shmatikov 2008) that exploit side information, and demonstrated their effectiveness on the Netflix prize dataset, an anonymized database contained viewing histories from customers of the movie streaming service. Other examples include attacks on shopping recommendation systems (Calandrino et al. 2011), or the identification of individuals from simple pictures taken in the street and linked with those publicly accessible on Facebook, where individuals are often identified by their friends or automatically by face recognition systems (Acquisti et al. 2014).

One type of data that is particularly difficult to anonymize is location data, such as mobility traces collected by smartphones. For example, de Montjoye et al. (2013) shows that knowing the four locations most frequently visited by a person gives a 95% chance to identify that person's trajectory in a large database of anonymous location traces. Moreover that percentage is still 50% if just two locations are known, which are typically a person's home and work address. In fact, coarsening the spatio-temporal sampling accuracy does not seem to decrease the ability to uniquely iden-tify trajectories before reaching levels at which the location data becomes of little utility (Zhang and Bolot 2011). These insights are particularly important as many researchers and city planners hope to get insight on human mobility from the analy-sis of datasets containing massive amounts of location traces. For example, the New York City Taxi and Limousine Commission released a dataset detailing millions of taxi trips, following a flawed anonymization process that clearly showed a lack of understanding of privacy issues (Tockar 2014).

The above examples of attacks are mostly re-identification attacks performed on databases publishing *microdata*, i.e., data at the level of individuals. Indeed, it appears that publishing microdata while providing meaningful privacy guarantees is in many cases very difficult, and privacy attacks on this type of data are the easiest to carry out. On the other hand, large-scale monitoring and control systems typically only require aggregate statistics about a population. These aggregate statistics could be computed from personal data streams, such as a dynamic map showing road traffic conditions built from location traces sent by smartphones, or an estimate of power consumption in a neighborhood updated using smart meter data from individual homes. Certain types of sensors also report only aggregated data, for example vehicle detectors on a road provide counts of vehicles passing at a certain location over a certain time period, e.g., one minute. Aggregation is beneficial to privacy, but examples have also shown that this is not sufficient to a priori rule out the possibility of significant privacy breaches. For example, individual trajectories can often be reconstructed from aggregate location data (Wilson and Atkeson 2005; Xu et al. 2017; Pyrgelis et al. 2017). Note also that re-identification attacks are not the only issue and other types of inferences about individual records in a database can be problematic and possibly easier to perform. Indeed, as noted before, simply identifying that someone's record is in a given database can be considered a significant breach of privacy, depending on the nature of the database, even if one cannot identify the specific record.

Privacy attacks and privacy preserving algorithms can be relevant in contexts that do not necessarily involve people. For example, firms might want to interact with each other or meet public reporting requirements while trying to hide proprietary information. A concrete example of such a situation arises in electricity markets, where inverse optimization techniques can be used to identify the cost functions of energy producers, using only the information published about daily market outcomes setting the electricity prices (Ruiz et al. 2013).

1.3 Differential Privacy

Our main goal in this monograph is to provide methods to compute aggregate statistics in real-time with formal privacy guarantees for the individuals from whom the data originates. First, we must consider the problem of precisely defining what we mean by preserving privacy. We wish to rely on a formal privacy metric, so that different solutions for a given data analysis problem can then be objectively compared both in terms of their accuracy and of the level of privacy they provide. Even though several approaches to privacy-preserving data analysis have been proposed, some of which are briefly discussed in Sect. 1.4, one particularly successful methodology, on which we focus here, relies on the notion of differential privacy.

1.3.1 Standard Definition

The notion of differential privacy is motivated by the following remark. When publishing the result of a useful statistical analysis based on a dataset, new information is released, *even potentially about individuals who did not provide their data*. To paraphrase an example in Dwork and Roth (2014), suppose that a new study analyzing a medical database links smoking to a higher risk of developing cancer. Publishing this result provides new information about all people known to be smokers, not just those whose record was in the dataset. Since we obviously do not want to rule out such analyses, what differential privacy aims for is not to prevent information disclosure per se, but to guarantee that if an individual provides her data, it does not become significantly easier to make new inferences about that specific individual compared to the situation where her records is not in the dataset. This can include preventing the detection that a given individual's record is in the dataset, since in the above scenario for example it might expose confidential information about her smoking habits. In summary, by publishing a differentially private statistic, the information disclosure about specific individuals should be essentially the same, whether or not that individual's record was in the dataset.

To achieve the desired outcome, a differentially private mechanism does not typically try to publish a whole dataset, even after perturbation, but only provides randomized responses to specific queries about this dataset from data analysts. It guarantees that the distribution over its published responses remains essentially the same even if the input dataset is changed to slightly different one, the typical situation being to change the dataset by adding, modifying or removing the record of a single individual. It is this "differential" point of view that allows the guarantee provided to hold against arbitrary adversaries, no matter what side information they have access to (Kasiviswanathan and Smith 2008).

Formally, we start by defining on the space U of datasets of interest a symmetric binary relation called adjacency and denoted Adj, which captures what it means for two datasets to differ by the data of a single individual. Essentially, it is hard to determine from a differentially private output which of any two adjacent input datasets was used to produce that output. A *mechanism* is just a randomized map $M : \mathsf{U} \to \mathsf{Y}$, for some measurable output space $(\mathsf{Y}, \mathscr{Y})$, where $\mathscr{Y}$ denotes a σ-algebra, such that for any element $u \in \mathsf{U}$, $M(u)$ is a random variable.[1] A mechanism can be viewed as a probabilistic algorithm to answer a given query q, which is simply a map $q : \mathsf{U} \to \mathsf{Y}$. In some cases, we index the mechanism by the query q of interest, writing M_q.

Example 1.1 Let $\mathsf{U} = \mathbb{R}^n$, with each real-valued entry of $u \in \mathsf{U}$ corresponding to some sensitive information for an individual contributing her data, e.g., her salary. A data analyst would like to know the average of the entries of u, i.e., the query is $q : \mathsf{U} \to \mathbb{R}$ with $q(u) = \frac{1}{n} \sum_{i=1}^{n} u_i$. As detailed in Chap. 2, a basic mechanism M_q to answer this query in a differentially private way computes $q(u)$ and blurs the result by adding the realization of a random variable Z (i.e., noise), so that $M_q(u) = \frac{1}{n} \sum_{i=1}^{n} u_i + Z$. Note that in the absence of perturbation Z, an adversary who knows n and all u_j for $j \geq 2$ can recover the remaining entry u_1 exactly if she/he learns $q(u)$. This can deter people from contributing their data, whereas we want to encourage a broader participation (increase n) to improve the accuracy of the analysis.

Our definition of a differentially private mechanism follows the original one proposed by Dwork et al. in their seminal paper (2006), simply generalizing here the notion of adjacency.

Definition 1.1 (*Differential privacy*) Let U be a space equipped with a symmetric binary relation denoted Adj, and let $(\mathsf{Y}, \mathscr{Y})$ be a measurable space. Let $\varepsilon \geq 0, \delta \in [0, 1]$. A mechanism M on U with values in Y is called (ε, δ)-differentially private for Adj if for all $u, u' \in \mathsf{U}$ such that Adj(u, u'), we have

---

[1] More formally, fix some probability space $(\Omega, \mathscr{F}, \mathbb{P})$. Then a mechanism is a map $M : \mathsf{U} \times \Omega \to \mathsf{Y}$, and we abbreviate the notation $M(u, \omega)$ by $M(u)$, following the standard practice used to denote random variables.

$$\mathbb{P}(M(u) \in S) \le e^{\varepsilon}\mathbb{P}(M(u') \in S) + \delta, \quad \forall S \in \mathscr{Y}. \tag{1.1}$$

If $\delta = 0$, the mechanism is simply said to be ε-differentially private.

In words, Definition 1.1 says that for two adjacent datasets, the distributions over the outputs of the mechanism should be close. The guarantee depends on the choice of adjacency relation, which can vary based on the application. The choice of the parameters ε, δ is set by the privacy policy. Typically ε is taken to be a small constant, e.g., $\varepsilon \approx 0.1$ or perhaps even $\ln 2$ or $\ln 3$. The parameter δ should be kept (very) small, as it controls the probability of certain significant losses of privacy. In particular, $\delta > 0$ in (1.1) allows a zero probability event for input u' to have positive probability for input u. Hence, by observing that the output is in the set S corresponding to this event, one would be able to determine with certainty that the input was u and not u'. Let us remark also here that because Adj is symmetric, the relation (1.1) automatically holds also for u and u' interchanged.

We should also note here that the notion of differential privacy in Definition 1.1 depends on the choice of σ-algebra $\mathscr{Y}$ on Y. Indeed, if $\mathscr{Y} = \{\emptyset, \mathsf{Y}\}$, then (1.1) becomes trivial, and so one should take a sufficiently rich class of measurable sets. In practice, either Y is countable and $\mathscr{Y}$ contains all subsets of Y, or Y is a metric space and $\mathscr{Y}$ is the Borel σ-algebra on Y.

1.3.2 Definition via Hypothesis Testing

The interpretation of differential privacy in terms of difficulty of distinguishing between adjacent datasets, mentioned earlier in this section, provides an alternative characterization of differentially private mechanisms, equivalent to Definition 1.1 (Wasserman and Zhou 2010; Kairouz et al. 2017). Namely, consider the problem of designing a statistical test to decide if an observed output $y \in \mathsf{Y}$ produced by a mechanism was computed based on a dataset u or an adjacent one u'. This test corresponds to choosing a set $S \subset \mathsf{Y}$ such that if y falls in S, we decide that the input dataset was u and if y is not in S, we decide that the input dataset was u'. Once S is chosen and y is observed, we can make two types of errors: deciding that the dataset is u when in fact it was u', which happens with probability $P_1 = \mathbb{P}(M(u') \in S)$, or deciding that the dataset is u' when in fact it was u, which happens with probability $P_2 = \mathbb{P}(M(u) \notin S)$. It turns out that M is differentially private if and only if we cannot make the probabilities of these two types of errors simultaneously small.

Theorem 1.1 (Wasserman and Zhou 2010; Kairouz et al. 2017) *Let $\varepsilon \ge 0$, $\delta \in [0, 1]$. Let U be a space equipped with an adjacency relation Adj. A mechanism M on U with values in $(\mathsf{Y}, \mathscr{Y})$ is (ε, δ)-differentially private for Adj if and only if for all Adj(u, u') and all sets $S \in \mathscr{Y}$, we have*

$$\mathbb{P}(M(u') \in S) + e^{\varepsilon}\mathbb{P}(M(u) \notin S) \ge 1 - \delta$$
$$and \quad e^{\varepsilon}\mathbb{P}(M(u') \in S) + \mathbb{P}(M(u) \notin S) \ge 1 - \delta. \tag{1.2}$$

The linear inequalities of Theorem 1.1 quantify the limits on the performance achievable by any statistical test to discriminate between u and u' based on just observing $M(u)$ (or $M(u')$). Since the condition is both necessary and sufficient, we can view the differential privacy property as a way to prevent adversaries to distinguish decisively between adjacent datasets, and this idea in practice guides the choice of the adjacency relation. The relations (1.2) can also be used to set the desired values of the privacy parameters ε, δ. For example, summing the two inequalities, we get for the error propabilities

$$\frac{1}{2}(P_1 + P_2) \geq \frac{1-\delta}{1+e^\varepsilon}, \tag{1.3}$$

so that ε, δ can be directly related to a lower bound on the sum of the error probabilities of any statistical test used to discriminate between adjacent datasets. Note that if we can assume a uniform prior distribution on u, the left hand side of (1.3) is the overall probability of error of the test.

1.3.3 Basic Composition Theorem and Resilience to Post-processing

Although Definition 1.1 tells us what property a mechanism should satisfy to be considered differentially private, it is only in Chap. 2 that we will start describing some methods to design differentially private mechanisms. For now, the following theorem tells us that complex differentially private mechanisms can be built by composing simpler ones and bounds the degradation of the privacy parameters under composition.

Theorem 1.2 (Basic composition theorem, informal) *Let* U *be a space equipped with an adjacency relation Adj. Let* M_1 *be an* $(\varepsilon_1, \delta_1)$-*differentially private mechanism on* U *with values in* Y_1. *Let* M_2 *be a mechanism on* $\mathsf{Y}_1 \times \mathsf{U}$ *with values in* Y_2 *and such that, for any* $y_1 \in \mathsf{Y}_1$, $M_2(y_1, \cdot)$ *is* $(\varepsilon_2, \delta_2)$-*differentially private. Then the composed mechanism* $M(u) := (M_1(u), M_2(M_1(u), u))$ *on* U *with values in* $\mathsf{Y}_1 \times \mathsf{Y}_2$ *is* $(\varepsilon_1 + \varepsilon_2, \delta_1 + \delta_2)$-*differentially private.*

A proof of this result is provided in Appendix B. By immediate induction from Theorem 1.2, the ε, δ parameters add (in the worst case) when composing any finite number of mechanisms, although this is a generic result providing only an upper bound on these privacy parameters. More refined composition theorems exist, see Dwork and Roth (2014). Next, suppose that the mechanism M_2 in Theorem 1.2, when receiving an value $M_1(u)$ from M_1, does not in fact re-access the dataset u to produce its answer. In other words, for any $y_1 \in \mathsf{Y}_1$, $M_2(y_1, \cdot)$ is in fact 0-differentially private (effectively, M_2 is just a mechanism on Y_1). Then Theorem 1.2 says that if we post-process the result of the (ε, δ)-differentially private mechanism M_1 by some (possibly randomized) function M_2, the same (ε, δ)-differential privacy guarantee still holds for the composed mechanism $M(u) = M_2(M_1(u))$. This is called the *resilience to post-processing* property of differential privacy, and is in fact a property that should hold for any proper definition of privacy, since we cannot control what post-processing operations might be performed by an adversary once a result is released. As we will see in the next chapters, this resilience to post-processing property is also at the core of a useful methods to design differentially private mechanisms, where some known generic mechanism is used to produce a differentially private output, which is then further processed to improve its accuracy for a given processing task.

1.3.4 Variations on Differential Privacy

Although differential privacy is originally motivated by hiding the presence of an individual in a dataset, more abstractly, Definition 1.1 ensures that any pair of adjacent datasets is hard to distinguish based on observing a differentially private output, no matter what the choice and interpretation for the adjacency relation is. Thus, we can define adjacency between two datasets by removing all data related to a single individual from one to obtain the other, but this is sometimes too strong a privacy requirement in order to obtain useful results. We can then weaken this relation to allow for example only bounded variations in the data of one individual between adjacent datasets.[2] In this case, differential privacy does not necessarily masks the presence of an individual in the dataset, but provides indistinguishability between different possible values of an individual's data.

For example, for geolocation services, one might want to hide the position of any individual within a given bounded radius. For a dataset $d = \{p_1, \ldots, p_n\}$ consisting of the locations of n individuals, a possible adjacency relation could then be

$$\text{Adj}(d, d') \text{ iff } \exists i \in [n] \text{ s.t. } \|p_i - p_i'\| \leq \rho_i \text{ and } \forall j \neq i, p_j = p_j', \qquad (1.4)$$

i.e., we provide quasi-indistinguishability between positions within a distance ρ_i for any individual i. This crucial idea of indistinguishability between adjacent datasets underlying the notion of differential privacy can also be implemented in ways that differ somewhat from Definition 1.1. For example, rather than embedding the notion of distance between datasets in the adjacency relation as in (1.4), Chatzikokolakis et al. (2013) instead defines "d-privacy" between arbitrary datasets by the condition

$$\mathbb{P}(M(u) \in S) \leq e^{d(u,u')}\mathbb{P}(M(u') \in S), \quad \forall S \in \mathscr{Y},$$

where the inputs u, u' belong to a space U equipped with a metric d. This is a variation on ε-differential privacy (compare to Definition 1.1), which captures directly the fact that a mechanism should produce similar output distributions for datasets that are

---

[2]This is sometimes implicitly done in the literature by "normalizing the dataset", assuming a priori that each individual's numerical data belong to a known bounded set for example.

close in the sense of d. The metric d should be defined based on the application, just as for the adjacency relation.

Other possible variations of Definition 1.1 consider different definitions of "closeness" for the random variables $M(u)$ and $M(u')$ when u and u' are adjacent, since (1.1) might not necessarily be the most mathematically convenient notion of distance between the distributions of $M(u)$ and $M(u')$. We do not pursue this discussion here, and refer the reader for example to Dwork and Rothblum (2016), Bun and Steinke (2016). Overall, the ideas and algorithms presented in this monograph should be adaptable to these variations on the definition of differential privacy without too much difficulty.

1.4 Alternative Approaches to Privacy-Preserving Data Analysis

Although it is outside our scope to provide a detailed discussion of alternative approaches to privacy-preserving data analysis, a topic with a rich history, it is still useful to place differential privacy in a more general context. Given the size of the related literature, we can only provide a few pointers in this brief section.

1.4.1 Statistical Disclosure Limitation, K-anonymity, and Information-Theoretic Approaches

Traditionally, various types of "statistical disclosure limitation" (SDL) techniques (Duncan and Lambert 1986; Templ 2014) have been applied by statisticians and economists when publishing statistics computed from private microdata, such as census data. Indeed, protecting the confidentiality of survey respondents can be a legal requirement for statistical agencies, or it might simply be a necessary step to ensure honest responses. Processing sensitive static data stored in medical or insurance databases for example faces similar issues. SDL often focuses on releasing perturbed versions of entire datasets. The data analyst identifies sensitive variables to protect, and applies various perturbations to the corresponding data such as adding noise or performing micro-aggregation, avoiding in particular to release table entries that are valid for only a few individuals and could thus help re-identification. Various risk measures can be used (Templ 2014), including k-anonymity (Sweeney 2002). Some SDL techniques such as randomized response also provide certain differential privacy guarantees (Dwork and Roth 2014, Chap. 3). Still, in most cases the proposed techniques did not aim to achieve a formally defined notion of privacy, which is required to reason more rigorously about privacy-utility trade-offs.

More recently, several definitions of privacy have thus been proposed, including k-anonymity (Sweeney 2002) and its various extensions (Li et al. 2007), information-theoretic measures of privacy (Sankar et al. 2013; Venkitasubramaniam 2013), conditions based on observability (Xue et al. 2014; Manitara and Hadjicostis 2013), and differential privacy (Dwork et al. 2006; Dwork 2006). The notion of k-anonymity is aimed at countering re-identification attacks on microdata and constrains the publication of a dataset in such a way that each combination of values for the sensitive attributes occurs in at least k individual records. A different, information-theoretic approach such as Sankar et al. (2013) assumes that a dataset is a realization of a random vector containing sensitive attributes X_h to keep hidden and other correlated attributes X_r for which a perturbed version $\hat{X}_r$, to determine, can be published. Privacy-utility trade-offs can then be studied for example by maximizing the conditional entropy $H(X_h|\hat{X}_r, Z)$ of the sensitive attributes given the revealed information as well as all the side information Z available, subject to a bound on the expected distortion $\mathbb{E}[\rho(X_r, \hat{X}_r)]$ achieved by the publication scheme, where ρ is some distance function. A practical limitation of such approaches is to require known statistical models for all the sources of information, including the side information Z. In particular, new privacy attacks on a dataset can be enabled when new sources of side information become available, in which case Z does not even exist when $\hat{X}_r$ is computed.

1.4.2 Privacy, Secrecy and the Role of Cryptography

It is also useful to point out in this introduction the difference between the notion of *privacy* that we consider in this monograph and another notion that we will call *secrecy* for the sake of clarity, but which is also sometimes called privacy in the literature. We are concerned here with privacy preserving data analysis, where the result of some computations is shared with parties other than the initial owners of the data, and these parties could be adversarial. A typical example is the public release of a statistic computed from a dataset containing sensitive information, such as a medical database. Third parties have then access to this statistic, and are free to try to combine this information with any other available information to try to make additional inferences about the individuals present in the database.

This scenario should be contrasted with the classical scenarios considered in cryptography, where two users trusting each other wish to exchange a secret message while preventing potential eavesdroppers intercepting their communication to decipher it. In such scenarios, the goal is to have ideally no communication of information to the external world, i.e., outside of the small set of trusted parties. Traditional tools from cryptography such as shared and public key cryptography aim at creating visible outputs from the private information that contains as little information about the underlying message as possible, so that they ideally appear random. Such outputs are clearly useless for a data analyst explicitly wishing to learn some information based on an underlying dataset.

Cryptographic tools can still be useful to implement certain distributed protocols for privacy-preserving data analysis, see for example Shi et al. (2011). However, the traditional techniques from cryptography are not sufficient, since information about the dataset will by definition necessarily leak in the desired outputs, and this leakage will then have to be carefully controlled in order to limit the impact of potential privacy attacks.

1.5   Privacy-Preserving Analysis for Dynamic Data

The main current trend motivating this book is the increasing emphasis on systems processing *streams* of dynamic data collected from many sources around us, from smartphones, to surveillance cameras to smart meters and house thermostats. With such systems, the computation of useful statistics in real-time (estimates of road traffic or power consumption in a small neighborhood) subject to privacy constraints brings new challenges to the fields of signal processing, systems and control. Our goal in this monograph is to present algorithms that can be used to analyse these streaming data sources while enforcing differential privacy guarantees.

References

Ackerman L (2013) Mobile health and fitness applications and information privacy: report to California consumer protection foundation. https://www.privacyrights.org/mobile-medical-appsprivacy-consumer-report.pdf

Acquisti A, Gross R, Stutzman F (2014) Face recognition and privacy in the age of augmented reality. J Priv Confid 6(2):1–20

Bun M, Steinke T (2016) Concentrated differential privacy: simplifications, extensions, and lower bounds. https://arxiv.org/abs/1605.02065

Calandrino JA et al (2011) You might also like: privacy risks of collaborative filtering. In: Proceedings of the IEEE symposium on security and privacy, Berkeley, CA

Chatzikokolakis K et al (2013) Broadening the scope of differential privacy using metrics. In: Proceedings of the 13th privacy enhancing technologies symposium, Bloomington, Indiana

de Montjoye Y-A et al (2013) Unique in the crowd: the privacy bounds of human mobility. Sci Rep 3:1376

Duncan G, Lambert D (1986) Disclosure-limited data dissemination. J Am Stat Assoc 81(393):10–28

Dwork C (2006) Differential privacy. In: Proceedings of the 33rd international colloquium on automata, languages and programming (ICALP), vol 4052. Lecture notes in computer science, Venice, Italy

Dwork C, Roth A (2014) The algorithmic foundations of differential privacy. Found Trends Theor Comput Sci 9(3–4):211–407

Dwork C, Rothblum GN (2016) Concentrated differential privacy. https://arxiv.org/abs/1603.01887

Dwork C et al (2006) Calibrating noise to sensitivity in private data analysis. In: Proceedings of the third theory of cryptography conference, New York, NY, pp 265–284

Electronic Privacy Information Center (EPIC) (2019). http://epic.org/

Federal Trade Commission (2014) Consumer generated and controlled health data. Spring privacy series, p 22

Kairouz P, Oh S, Viswanath P (2017) The composition theorem for differential privacy. IEEE Trans Inf Theory 63(6):4037–4049

Kasiviswanathan SP, Smith A (2008) A note on differential privacy: defining resistance to arbitrary side information. arXiv:0803.3946

Li N, Li T, Venkatasubramanian S (2007) t-closeness: privacy beyond kanonymity and l-diversity. In: Proceedings of the 23rd IEEE international conference on data engineering

Manitara NE, Hadjicostis CN (2013) Privacy-preserving asymptotic average consensus. In: Proceedings of the European control conference, Zurich, Switzerland

Markey EJ (2015) Tracking and Hacking Security and Privacy Gaps Put American Drivers at Risk. Technical report U.S. Senator's report

McDaniel P, McLaughlin S (2009) Security and privacy challenges in the smart grid. IEEE Secur Priv 7(3):75–77

Narayanan A, Shmatikov V (2008) Robust de-anonymization of large sparse datasets (how to break anonymity of the netflix prize dataset). In: Proceedings of the IEEE symposium on security and privacy, Oakland, CA

President's Council of Advisors on Science and Technology (2016) Big data and privacy: a technological perspective. Technical report, Executive Office of the President of the United States

Pyrgelis A, Troncoso C, De Cristofaro E (2017) What does the crowd say about you? evaluating aggregation-based location privacy. In: Proceedings on privacy enhancing technologies

Ruiz C, Conejo AJ, Bertsimas DJ (2013) Revealing rival marginal offer prices via inverse optimization. IEEE Trans Power Syst 28(3):3056–3064

Sankar L, Rajagopalan SR, Poor HV (2013) Utility-privacy tradeoffs in databases: an information-theoretic approach. IEEE Trans Inf Forensics Secur 8(6):838–852

Shi E et al (2011) Privacy-preserving aggregation of time-series data. In: Proceedings of 18th annual network and distributed system security symposium (NDSS 2011)

Sweeney L (2002) k-anonymity: a model for protecting privacy. Int J Uncertain Fuzziness Knowl-Based Syst 10(05):557–570

Sweeney L (1997) Weaving technology and policy together to maintain confidentiality. J Law Med Ethics 25:98–110

Templ M et al (2014) Introduction to statistical disclosure control (SDC). Technical report, International Household Survey Network

Tockar A (2014) Riding with the stars: passenger privacy in the NYC taxicab dataset. Neustar Research. https://research.neustar.biz/2014/09/15/riding-with-the-stars-passenger-privacy-in-the-nyc-taxicab-dataset/

Venkitasubramaniam P (2013) Decision making under privacy restrictions. In: IEEE Conference on Decision and Control

Warren SD, Brandeis LD (1890) The right to privacy. Harv Law Rev 4(5):193–220. https://doi.org/10.2307/1321160

Wasserman L, Zhou S (2010) A statistical framework for differential privacy. J Am Stat Assoc 105(489):375–389

Who's reading your medical records? (1994) Consumer Reports. October 1994, pp 628–632

Wilson DH, Atkeson C (2005) Simultaneous tracking and activity recognition (STAR) using many anonymous, binary sensors. In: Gellersen H-W, Want R, Schmidt A (eds) Pervasive computing, vol 3468. Lecture notes in computer science. Springer, Berlin, pp 62–79

Xu F et al (2017) Trajectory recovery from ash: user privacy is not preserved in aggregated mobility data. In: Proceedings of the 26th international conference on world wide web, pp 1241–1250

Xue M, Wang W, Roy S (2014) Security concepts for the dynamics of autonomous vehicle networks. Automatica 50:852–857

Zhang H, Bolot J (2011) Anonymization of location data does not work: a large-scale measurement study. In: Proceedings of the 17th annual international conference on mobile computing and networking

Chapter 2
Basic Differentially Private Mechanisms

2.1 Introduction

[1]The Definition 1.1 of differential privacy specifies what guarantees we would like our algorithms to enforce, but not specific methods that can be used to achieve this goal. The purpose of this chapter is to present some basic tools to design mechanisms able to release potentially infinite data streams with differential privacy guarantees. Since our main focus in this monograph is on real-valued signals, we start with the problem of publishing finite sequences of real values, i.e., vectors, and introduce for this purpose the Laplace and Gaussian mechanisms, providing ε and (ε, δ)-differential privacy guarantees respectively.

2.2 Laplace and Gaussian Mechanisms

A vector-valued query on a space U (containing our datasets) is simply a function $q : \mathsf{U} \to \mathsf{V}$, for some vector space V, assumed equipped with a norm. For example, we can have $\mathsf{V} = \mathbb{R}^d$ for some positive finite integer d, equipped with the p-norm for some $p \in [1, \infty)$, defined for all $x = [x_1, \ldots, x_d]^T \in \mathbb{R}^d$ as

$$|x|_p = \left(\sum_{i=1}^{d} |x_i|^p \right)^{1/p}.$$

Of particular interest in this book is the case where V is the space of discrete-time (one-sided) sequences $x := \{x_t\}_{t \geq 0}$ with $x_t \in \mathbb{R}^d$ for all $t \geq 0$, a vector space that we denote generically $\mathsf{S}^d := (\mathbb{R}^d)^{\mathbb{N}}$. We can equip it with an ℓ_p-norm

---

[1]Some of the text in Sect. 2.3.3.3 of this chapter is reprinted, with permission, from Le Ny and Mohammady (2018) (©[2018] IEEE).

$$\|x\|_p = \left(\sum_{t=0}^{\infty} |x_t|_p^p \right)^{1/p}, \tag{2.1}$$

for some $p \in [1, \infty)$. Note that in (2.1) we use the p-norm on $\mathbb{R}^d$ for the same value of p. The vector space of sequences x in S^d such that $\|x\|_p < \infty$ is denoted ℓ_p^d.

For differential privacy, we need to introduce an adjacency relation Adj on U, as explained in Chap. 1. Specific adjacency relations on S^d are discussed later in Sect. 2.3. For now, we introduce the notion of sensitivity for a given adjacency relation, which plays a major role in the design of differentially private algorithms.

Definition 2.1 (*Sensitivity*) Let U be a space equipped with an binary symmetric adjacency relation Adj. Let V be a vector space equipped with a norm $\| \cdot \|$. The sensitivity of a query $q : \mathsf{U} \to V$ (for Adj and $\| \cdot \|$) is defined as

$$\Delta q = \sup_{\mathrm{Adj}(u,u')} \|q(u) - q(u')\|.$$

In particular, if V is $\mathbb{R}^d$, for some positive integer d, equipped with the p-norm, or if $V = \ell_p^d$ equipped with the ℓ_p-norm, we call Δq the p- or ℓ_p-sensitivity respectively and we use the notation $\Delta_p q$.

Given a vector valued query q, it turns out that by adding noise with a Laplace distribution and standard deviation proportional to the 1-sensitivity of the query, or with a Gaussian distribution proportionally to the 2-sensitivity, one can obtain ε- and (ε, δ)-differentially private mechanisms respectively. First, to introduce the Laplace mechanism (Dwork et al. 2006a), recall that the centered Laplace probability density function (pdf) is $\exp(-|x|/b)/2b$, with corresponding variance $2b^2$. The notation $X \sim \mathrm{Lap}^d(b)$ means that a random vector X has d independent and identically distributed (iid) components all following a centered Laplace distribution with the same parameter b. The pdf of X on $\mathbb{R}^d$ is then $\exp(-|x|_1/b)/(2b)^d$.

Theorem 2.1 (Laplace mechanism) *Let $\varepsilon > 0$ and fix an adjacency relation Adj on some set U. For a vector-valued query $q : \mathsf{U} \to \mathbb{R}^d$ with 1-sensitivity $\Delta_1 q$, the Laplace (additive noise) mechanism $M(u) = q(u) + w$, with $w \sim \mathrm{Lap}^d(b)$ and $b \geq \Delta_1 q / \varepsilon$, is ε-differentially private for Adj.*

Proof Consider two adjacent inputs u and u'. We have, for $S \subset \mathbb{R}^d$ measurable,

$$\begin{aligned}
P(M(u) \in S) &= \left(\frac{1}{2b} \right)^d \int_{\mathbb{R}^d} 1_S(q(u) + w) e^{-|w|_1/b} dw \\
&= \left(\frac{1}{2b} \right)^d \int_{\mathbb{R}^d} 1_S(z) e^{-|z - q(u)|_1/b} dz \\
&\leq e^{|q(u) - q(u')|_1/b} \left(\frac{1}{2b} \right)^d \int_{\mathbb{R}^k} 1_S(z) e^{-|z - q(u')|_1/b} dz \\
&= e^{|q(u) - q(u')|_1/b} P(M(u') \in S),
\end{aligned}$$

where the inequality comes from $-|z - q(u)|_1 \leq -|z - q(u')|_1 + |q(u) - q(u')|_1$ by the triangle inequality. Since $|q(u) - q(u')|_1 \leq \Delta_1 q$, with the choice $b \geq \Delta_1 q / \varepsilon$, we obtain the inequality (1.1) certifying ε-differential privacy (with $\delta = 0$). $\square$

Remark 2.1 The proof of Theorem 2.1 in fact shows that a mechanism publishing $M(u) = q(u) + w$ with w a random vector in $\mathbb{R}^d$ generated using a probability density function of the form

$$f(w) = K \exp\left(-\frac{\|w\| \varepsilon}{\Delta q}\right),$$

with K a normalizing constant, is ε-differentially private, for *any* choice of norm $\|\cdot\|$, provided the sensitivity Δq is computed using this norm. For example, we could take the p-norm for $p \geq 1$ and the corresponding p-sensitivity. Such schemes can be useful in situations where the p-sensitivity is smaller than the 1-sensitivity, for example the 2-sensitivity scales with the square root of the dimension d, whereas the 1-sensitivity scales linearly with d. However, the corresponding multivariate noise is more difficult to generate than the Laplace noise, since the components of w are no longer independent. More importantly for the topic of this monograph, the marginals do not follow a distribution of the same form, which makes such schemes apparently impractical for perturbing data streams in real-time, since in this case the dimension of the random vector essentially grows with each sample, as we discuss below in Sect. 2.3.

Next, we introduce the Gaussian mechanism (Dwork et al. 2006b). It is similar to the Laplace mechanism but adds iid Gaussian noise to provide (ε, δ)-differential privacy, with $\delta > 0$ but typically a smaller ε for the same accuracy. Recall first the definition of the Q-function $Q(x) := \frac{1}{\sqrt{2\pi}} \int_x^\infty e^{-u^2/2} du$, i.e., the tail distribution of the standard normal distribution. Since Q is monotonically decreasing, it has a decreasing inverse Q^{-1}, which maps the interval $(0, 1)$ into $(-\infty, \infty)$. For $\varepsilon > 0$, $1 > \delta > 0$, define the notation for the following necessarily positive quantity

$$\kappa_{\delta, \varepsilon} = \frac{1}{2\varepsilon}(Q^{-1}(\delta) + \sqrt{(Q^{-1}(\delta))^2 + 2\varepsilon}). \tag{2.2}$$

The notation $X \sim \mathcal{N}(\mu, \Sigma)$ means that the random vector X follows a multivariate Gaussian distribution with mean vector μ and covariance matrix Σ, and I_d is the $d \times d$ identity matrix.

Theorem 2.2 (Gaussian mechanism) *Let $\varepsilon > 0$, $1 > \delta > 0$, and fix an adjacency relation Adj on some set* U. *For a vector-valued query $q : \mathsf{U} \to \mathbb{R}^d$ with 2-sensitivity $\Delta_2 q$, the Gaussian (additive noise) mechanism $M(u) = q(u) + w$, with $w \sim \mathcal{N}\left(0, \sigma^2 I_d\right)$ and $\sigma \geq \Delta_2 q\, \kappa_{\delta, \varepsilon}$, is (ε, δ)-differentially private.*

Proof Consider two adjacent inputs u, u', and denote $v := q(u) - q(u')$. For $S \subset \mathbb{R}^d$ measurable, we have

$$\mathbb{P}(M(u) \in S) = \frac{1}{(2\pi\sigma^2)^{d/2}} \int_{\mathbb{R}^d} 1_S(q(u) + w) e^{-|w|_2^2/(2\sigma^2)} dw$$

$$= \frac{1}{(2\pi\sigma^2)^{d/2}} \int_{\mathbb{R}^d} 1_S(z) e^{-|z-q(u)|_2^2/(2\sigma^2)} dz$$

$$= \frac{1}{(2\pi\sigma^2)^{d/2}} \int_{\mathbb{R}^d} 1_S(z) e^{-|z-q(u')-v|_2^2/(2\sigma^2)} dz$$

$$= \frac{1}{(2\pi\sigma^2)^{d/2}} \int_S \exp\left(-\frac{|z-q(u')|_2^2}{2\sigma^2}\right) \exp\left(\frac{2(z-q(u'))^T v - |v|_2^2}{2\sigma^2}\right) dz$$

$$= \frac{1}{(2\pi\sigma^2)^{d/2}} \int_{S \cap A_\varepsilon} \exp\left(-\frac{|z-q(u')|_2^2}{2\sigma^2}\right) \exp\left(\frac{2(z-q(u'))^T v - |v|_2^2}{2\sigma^2}\right) dz$$

$$+ \frac{1}{(2\pi\sigma^2)^{d/2}} \int_{S \cap A_\varepsilon^c} \exp\left(-\frac{|z-q(u)|_2^2}{2\sigma^2}\right) dz,$$

where $A_\varepsilon = \left\{z \in \mathbb{R}^d : \frac{2(z-q(u'))^T v - |v|_2^2}{2\sigma^2} \leq \varepsilon\right\}$ and A_ε^c denotes its complement. By definition of A_ε, the first term of the last expression is bounded by

$$e^\varepsilon \frac{1}{(2\pi\sigma^2)^{d/2}} \int_S \exp\left(-\frac{|z-q(u')|_2^2}{2\sigma^2}\right) dz = e^\varepsilon \mathbb{P}(M(u') \in S).$$

The second integral term is bounded by

$$\frac{1}{(2\pi\sigma^2)^{d/2}} \int_{\mathbb{R}^d} \exp\left(-\frac{|z-q(u)|_2^2}{2\sigma^2}\right) 1_{\{2(z-q(u'))^T v \geq |v|_2^2 + 2\varepsilon\sigma^2\}} dz,$$

which, after the change of variable $y = (z - q(u))/\sigma$, can be rewritten

$$\frac{1}{(2\pi)^{d/2}} \int_{\mathbb{R}^d} \exp\left(-\frac{|y|_2^2}{2}\right) 1_{\{2(\sigma y + v)^T v \geq |v|_2^2 + 2\varepsilon\sigma^2\}} dy$$

$$= \frac{1}{(2\pi)^{d/2}} \int_{\mathbb{R}^d} \exp\left(-\frac{|y|_2^2}{2}\right) 1_{\{y^T v \geq \varepsilon\sigma - |v|_2^2/2\sigma\}} dy.$$

This last expression is $\mathbb{P}\left(Y^T v \geq \varepsilon\sigma - \frac{|v|_2^2}{2\sigma}\right)$, for $Y \sim \mathcal{N}(0, I_d)$. In particular, $Y^T v \sim \mathcal{N}(0, |v|_2^2)$, hence is equal to $|v|_2 Z$ in distribution, with $Z \sim \mathcal{N}(0, 1)$. We are then led to set σ sufficiently large so that $\mathbb{P}(Z \geq \varepsilon\sigma/|v|_2 - |v|_2/2\sigma) \leq \delta$, i.e., $Q(\varepsilon\sigma/|v|_2 - |v|_2/2\sigma) \leq \delta$. Because Q^{-1} is monotonically decreasing, we must then have

$$\frac{\varepsilon\sigma}{|v|_2} - \frac{|v|_2}{2\sigma} \geq Q^{-1}(\delta), \text{ i.e., } \sigma^2 - \sigma\frac{|v|_2}{\varepsilon}Q^{-1}(\delta) - \frac{|v|_2^2}{2\varepsilon} \geq 0.$$

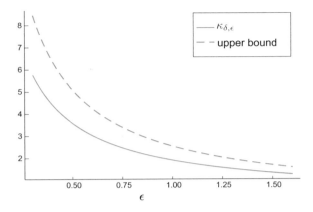

Fig. 2.1 A comparison of the Gaussian mechanism parameter $\kappa_{\delta,\varepsilon}$ given by (2.2) and its upper bound $\sqrt{2\ln(1.25/\delta)}/\varepsilon$, for $\delta = 0.05$

A straightforward analysis of the second-order equation shows that the inequality is satisfied when $\sigma \geq |v|_2 \, \kappa_{\delta,\varepsilon}$, and hence in particular for $\sigma \geq \Delta_2 q \, \kappa_{\delta,\varepsilon}$, since $|v|_2 \leq \Delta_2 q$ when u and u' are adjacent. □

As an illustration of Theorem 2.2, to guarantee (ε, δ)-differential privacy with $\varepsilon = \ln 2$ and $\delta = 0.05$, the standard deviation of the Gaussian noise should be about $\kappa_{\delta,\varepsilon} = 2.65$ times the 2-sensitivity of q. It can be shown that $\kappa_{\delta,\varepsilon}$ can be bounded by $\sqrt{2\ln(1.25/\delta)}/\varepsilon$ (Dwork and Roth 2014). This clarifies intuitively the scaling of the noise's standard deviation as δ, ε become small, but since the bound is not tight, it is preferable to use (2.2) in implementations (for the values of $\delta = 0.05$, $\varepsilon = \ln 2$ above for example, the bound gives approximately 3.53). A comparison of $\kappa_{\delta,\varepsilon}$ with its bound is shown on Fig. 2.1.

To conclude this section, we extend the Laplace and Gaussian mechanisms to signal-valued queries. Differentially private sequences in S^d can be obtained by adding white noise, i.e., a sequence of iid zero-mean random variables, to a desired output. The main change is to replace the p-norm on $\mathbb{R}^d$ by the ℓ_p-norm on S^d to compute the sensitivity.

Theorem 2.3 (Laplace and Gaussian mechanisms for signal-valued queries) *Let $\varepsilon > 0, 1 > \delta > 0$, and fix an adjacency relation Adj on some set U. Consider a query $q : \mathsf{U} \to S^d$ with ℓ_p-sensitivity $\Delta_p q$ for $p = 1, 2$. The Laplace mechanism $M(u) = q(u) + w$, with $w = \{w_t\}_{t \geq 0}$ a sequence of iid random variables such that $w_t \sim Lap^d(b_t)$ and $b_t \geq \Delta_1 q/\varepsilon$ for $t \geq 0$, is ε-differentially private for Adj. The Gaussian mechanism, where instead $w_t \sim \mathcal{N}\left(0, \sigma_t^2 I_d\right)$ and $\sigma_t \geq \Delta_2 q \, \kappa_{\delta,\varepsilon}$, is (ε, δ)-differentially private.*

Proof The proof, which can be found in Le Ny and Pappas (2014), consists mainly in resolving technical measurability issues for infinite dimensional signal spaces when extending Theorems 2.1 and 2.2 to infinite sequences. □

2.3   Input and Output Mechanisms: Basic Examples

We present and compare in this section two simple differentially private mechanisms to publish numerical data, either finite-dimensional vectors or infinite vector-valued sequences. The first mechanism adds noise directly to the sensitive data and is called the input perturbation mechanism. The second mechanism adds noise to the result of the query computation, and is called the output perturbation mechanism. To implement the output perturbation mechanism, one needs to be able to either compute the query's sensitivity exactly or obtain a tight upper bound. In general, this requires choosing the adjacency relation carefully, in order to both capture the privacy constraint appropriately and allow the computation of sufficiently tight sensitivity bounds. Hence, this section also serves to introduce examples of adjacency relations and systems for which the sensitivity can be easily computed.

2.3.1   Adjacency Relations on Individual and Collective Data Streams

To discuss more concretely adjacency relations on signals, it is useful to distinguish between two types of data streams, which we call "individual" and "collective" data streams in the following. These correspond to two types of sensing modalities. *Individual* signals are simply signals related to single individuals. An individual signal could come for example from a GPS device carried by this individual, transmitting and updating in real-time to a third party a location history for this individual. Examples include Google's location history service, and probe vehicle data used for traffic estimation. On the other hand, *collective* signals or data streams are not directly associated with a single individual but rather with a group of individuals, and hence already include some form of aggregation. For example, traffic sensors placed at fixed positions such as induction loops, electronic toll pass readers and cameras provide counts specifying the number of cars passing in front of a sensor during a given interval, rather than track individual cars along the road. Similarly, motion detectors in buildings are triggered by any person passing close to them. For another example, an electricity meter for a house provides the total power consumption of all individual appliances in the house, which is a collective signal for a situation where the individual data we might want to protect could be the consumption from a single appliance rather than a single person.

It is important to note that although some form of aggregation is present in collective signals, it is not by itself a guarantee of privacy. For example, it is well-known that power consumption traces can be "disaggregated" by leveraging knowledge about the power consumption signature of individual appliances, and as a result detailed information about the activities of a household can be inferred from electricity consumption traces (Hart 1992; Molina-Markham et al. 2010). The key, as usual, is the possibility of linking published data with other sources of information. A motion sensor placed right outside of an individual's office will be triggered in particular every time the person leaves or returns to his or her office, thereby leaking some information about that specific individual's activities, and could therefore be intuitively more privacy-intrusive than the same sensor placed in a common meeting area. See also Xu et al. (2017), Pyrgelis et al. (2017) for examples of privacy attacks on aggregated location data.

In general, we want to tailor the adjacency definition to the application and in particular to the fact that the raw privacy-sensitive data consists of individual or collective signals. With individual signals, we are interested in systems that take possibly many such signals as inputs and produce one or a few aggregate output signals based on these inputs. Such a system could then be a mapping $\mathcal{G} : \mathsf{S}_1 \times \cdots \times \mathsf{S}_n \to \mathsf{S}^m$, with n the number of individual signals, $\mathsf{S}_i = \mathsf{S}^{d_i}$ for some integer d_i, m the number of outputs, and in general $n \gg m$. For differential privacy, we would like to make it hard to detect certain variations in any input from observing just the output. Hence, a possible adjacency relation on $\mathsf{S}_1 \times \cdots \times \mathsf{S}_n$, assuming each S_i is equipped with a norm $\| \cdot \|_p$, could be, for some positive constant ρ,

$$\mathrm{Adj}(u, u') \text{ iff } \exists i \in [n] \text{ s.t. } \|u_i - u'_i\|_p < \rho \text{ and } u_j = u'_j, \forall j \neq i. \qquad (2.3)$$

That is, since each component of the multidimensional input signal is associated with a unique individual, two input signals u and u' are considered adjacent if they differ in only one of their components. Moreover, we bound here the allowed variations in the pairs of adjacent input signals to obtain finite sensitivity bounds.[2]

Whereas individuals directly produce personal signals, they only have a limited effect on collective signals. For example, a single car traveling in a given direction on a highway affects the counts of a detector at a fixed position by at most one. Of course, this car can be detected over time by many sensors distributed in many places, and from this data it might actually be possible to track it. But for the collective *scalar* signals coming from a single fixed sensor, we could have an adjacency relation of the form

$$\mathrm{Adj}(u, u') \text{ iff } \exists t_0 \text{ s.t. } |u_{t_0} - u'_{t_0}| \leq \rho \text{ and } u_t = u'_t, \forall t \neq t_0, \qquad (2.4)$$

for some fixed constant $\rho > 0$. For m collective signals, e.g., coming from m fixed sensors, and given a vector of positive constants $\rho = [\rho_1, \ldots, \rho_m] \in \mathbb{R}^m_{>0}$, this adjacency relation could be generalized to

$$\mathrm{Adj}(u, u') \text{ iff } \exists t_1, \ldots, t_m \text{ s.t. } \forall 1 \leq i \leq m, |u_{i,t_i} - u'_{i,t_i}| \leq \rho_i \text{ and } u_{i,t} = u'_{i,t}, \forall t \neq t_i. \qquad (2.5)$$

---

[2]It is possible to generalize the discussion to norms p_i, bounds ρ_i and even privacy levels ε_i, δ_i that depend on each individual i. The extension of the results presented in the following to this more general set-up follows similar arguments.

In both cases, the interpretation is that a single individual has a (bounded) effect on a given input data stream only at a single time-period, at least during the time frame of interest for the analysis. Imagine for example an individual using a smartphone app to "check-in" at various stores during a day. A single data stream originates from each store, and presumably a single person would only check-in once during a day at a particular location. Differential privacy with respect to (2.5) then makes it hard to infer if a person visited a given store, as long as that person visited the store only once over the period of interest for the analysis. Handling multiple successive visits could be done by "virtually" duplicating sensor i for different time windows, over which only one visit is allowed.

The adjacency relation (2.4) was introduced in the differential privacy literature in Dwork et al. (2010), under the name "event-level privacy", and was later also used in Chan et al. (2011), Bolot et al. (2013) for example. The relation (2.5) is a generalization to multidimensional signals. Another possible generalization of interest is the following

$$\text{Adj}(u, u') \text{ iff } \exists t_0, \alpha, \text{ s.t. } \begin{cases} u_t = u'_t, t < t_0, \\ |u_t - u'_t|_p \leq B_p\, \alpha^{t-t_0}, t \geq t_0, \end{cases} \tag{2.6}$$

where $| \cdot |_p$ is the p-norm on $\mathbb{R}^d$ for some integer $p \in [1, \infty)$ and $B_p > 0, 1 > \alpha \geq 0$ are constants. In this case, the influence of a single individual on the (possibly multidimensional) input signal is allowed to last more than one period (unless $\alpha = 0$), as long as it decreases geometrically. Adjacency definitions such as (2.6) are motivated both by trying to provide stronger privacy guarantees (by relaxing the adjacency constraint as much as possible, to protect against more variations) while still being able to compute tight bounds on sensitivity, thereby keeping the privacy-preserving noise to a minimum. A further relaxation of the adjacency condition (2.6) is to take simply

$$\text{Adj}(u, u') \text{ iff } \|u - u'\|_p \leq B_p, \tag{2.7}$$

for some ℓ_p-norm and some positive constant B_p. This adjacency relation then leads to calculations that are similar to those for (2.3). It is weaker than (2.6), i.e., admits more adjacent signals, hence provides a stronger privacy guarantee for given values of (ε, δ), although of course one might then need a higher level of noise to enforce differential privacy with respect to this adjacency relation.

2.3.2 Input Perturbation Mechanisms

Input perturbation mechanisms simply add privacy-preserving noise to the input signals. Any subsequent processing of these private signals still produces private signals, by the resilience to post-processing property discussed in Sect. 1.3.3. When these input signals are individual signals, data providers can add the noise themselves and therefore directly release differentially private signals. Such a scheme has the advantage of not requiring the presence of a trusted data aggregator to release private results, but one obtains meaningful answers to queries only when there is sufficient "averaging" of the input noise. Input perturbation Laplace and Gaussian mechanisms based on Theorem 2.3 are straightforward to implement, provided we have chosen the adjacency relation in such a way that the ℓ_1 and/or ℓ_2-sensitivity of the identity operator is easy to compute.

Corollary 2.1 *Let u be a sequence in $S_1 \times \cdots \times S_n$ and consider the perturbed sequence $v_t = u_t + w_t$ for w a white noise signal. Then, for the adjacency relation (2.3) with $p = 1$, the signal v is ε-differentially private if $w_t \sim Lap\,(b/\varepsilon)^d$ (with $d = \sum_{i=1}^n d_i$) for $b \geq \rho$. It is (ε, δ)-differentially private if $p = 2$ in (2.3) and $w_t \sim \mathcal{N}(0, \sigma^2 I_d)$ for $\sigma \geq \kappa_{\delta,\varepsilon}\rho$.*

Note that in Corollary 2.1 the main reason for using Gaussian rather than Laplace noise is that we relax the constraint in the adjacency relation by changing the ℓ_1-norm to the ℓ_2-norm. Indeed, since we always have $\|x\|_2 \leq \|x\|_1$ for a signal x (and the gap can be arbitrarily large, take for example $x_k = 1/(k+1)$), for a given constant ρ and a given signal u, there are more signals u' that are adjacent to u according to (2.3) when $p = 2$ than when $p = 1$, and hence more variations in a user input for which we can provide a differential privacy guarantee, albeit with $\delta > 0$. Let us now consider collective signals.

Corollary 2.2 *Let u be a sequence in S^m, and consider the perturbed sequence $v_t = u_t + w_t$ for w a white noise signal. Then, for the adjacency relation (2.5), the signal v is ε-differentially private if $w_t \sim Lap\,(b/\varepsilon)^m$, for $b \geq \|\rho\|_1$. It is (ε, δ)-differentially private if $w_t \sim \mathcal{N}(0, \sigma^2 I_m)$ for $\sigma \geq \kappa_{\delta,\varepsilon} \|\rho\|_2$.*

Proof For two adjacent signals u, u' according to (2.5), we have

$$u - u' = [z_1 \delta^{t_1}, \ldots, z_m \delta^{t_m}]^T, \tag{2.8}$$

for some $|z_i| \leq \rho_i$, with δ^{t_0} denoting a discrete delta signal at t_0, i.e., $\delta_t^{t_0} = 1$ if $t = t_0$ and 0 otherwise. In particular

$$\|u - u'\|_1 = \sum_{i=1}^m |z_i| \leq \|\rho\|_1, \text{ and similarly } \|u - u'\|_2 = \sqrt{\sum_{i=1}^m |z_i|^2} \leq \|\rho\|_2.$$

With these bounds on the sensitivity of the identity query, we apply Theorem 2.3. $\square$

As a final example, consider input perturbation mechanisms for the adjacency relation (2.6).

Corollary 2.3 *Let u be a sequence in S^d, and consider the perturbed sequence $v_t = u_t + w_t$ for w a white noise signal. Then, for the adjacency relation (2.6), the signal*

v is ε-differentially private if $p = 1$ and $w_t \sim Lap\,(b/\varepsilon)^d$, for $b \geq B_1/(1 - \alpha)$. It is (ε, δ)-differentially private if $p = 2$ and $w_t \sim \mathcal{N}(0, \sigma^2 I_d)$ for $\sigma \geq \kappa_{\delta,\varepsilon} B_2/\sqrt{1 - \alpha^2}$.

Proof The result follows simply by computing $\|u - u'\|_p$ for $p = 1$ and $p = 2$ when u and u' satisfy (2.6). We then apply Theorem 2.3. $\qquad\qquad\qquad\square$

2.3.3 Output Perturbation Mechanisms

Output perturbation mechanisms rely directly on Theorem 2.3 to add noise after the result of a desired query has been computed. For queries specified by linear time-invariant (LTI) systems, sensitivity can be computed using system norms, when an appropriate definition of adjacency is chosen. Thus, we start this section by recalling some key facts about system norms.

2.3.3.1 Brief Review of System Norms

For each time index T, let P_T be the truncation operator, so that for any signal x

$$(P_T x)_t = \begin{cases} x_t, & t \leq T, \\ 0, & t > T. \end{cases} \tag{2.9}$$

A deterministic system $\mathscr{G} : \mathsf{S}^m \to \mathsf{S}^n$ is causal if and only if $P_T \mathscr{G} = P_T \mathscr{G} P_T$, in other words, the output up to time T is not influenced by the input at times $t > T$. We denote by ℓ_{pe}^d the extended ℓ_p^d-space, i.e., the space of sequences with values in $\mathbb{R}^d$ and such that $x \in \ell_{pe}^d$ if and only if $P_T x$ has finite ℓ_p-norm for all integers T. Consider now a causal discrete time LTI system $\mathscr{G}$ with matrix-valued transfer function $z \mapsto G(z)$ and impulse response $\{G_t\}_{t \geq 0}$, so that $G(z) = \sum_{t=0}^{\infty} G_t z^{-t}$. The H_∞ norm of this system $\mathscr{G}$ (or of the transfer matrix G) is

$$\|G\|_\infty = \operatorname*{ess\,sup}_{\omega \in (-\pi, \pi]} \|G\left(e^{j\omega}\right)\|_2,$$

where $\|A\|_2 := \sigma_{\max}(A)$ denotes the maximum singular value of a matrix A. The H_2 norm of the system $\mathscr{G}$ (or of the transfer matrix G) is

$$\|G\|_2 = \frac{1}{2\pi} \int_{-\pi}^{\pi} \operatorname{Tr}\left(G\left(e^{j\omega}\right)^* G\left(e^{j\omega}\right)\right) d\omega = \frac{1}{2\pi} \int_{-\pi}^{\pi} \|G\left(e^{j\omega}\right)\|_F^2 d\omega = \sum_{t=0}^{\infty} \|G_t\|_F^2,$$

where $\|A\|_F = \sqrt{\operatorname{Tr}(A^* A)} = \sqrt{\sum_{i,j} |A_{ij}|^2}$ denotes the Frobenius norm of a (possibly complex-valued) matrix A, and the last equality is due to Parseval's identity.

Recall some basic interpretations of the H_∞ and H_2 norms. For a system $\mathscr{G}$ with inputs in ℓ_{re}^d and output in $\ell_{se}^{d'}$, its ℓ_r-to-ℓ_s incremental gain $\gamma_{r,s}(\mathscr{G})$ is defined as the smallest number γ such that

$$\|P_T \mathscr{G}u - P_T \mathscr{G}u'\|_s \leq \gamma \|P_T u - P_T u'\|_r, \quad \forall u, u' \in \ell_{re}^d, \ \forall T.$$

The following is a key result from system theory.

Theorem 2.4 *If $\mathscr{G}$ is a finite-dimensional LTI system with transfer matrix G, then $\gamma_{2,2}(\mathscr{G}) = \|G\|_\infty$.*

The H_2 norm has several motivations, two of which are of particular importance in the following. The first interpretation concerns stochastic inputs. We refer the reader to Appendix A for a brief review of basic notions of statistical signal processing.

Theorem 2.5 *Let w be a discrete-time wide-sense stationary (WSS) stochastic process with power spectral density (PSD) matrix $S_w\left(e^{j\omega}\right)$. Define $W(z)$ so that*

$$S_w\left(e^{j\omega}\right) = W\left(e^{j\omega}\right) W\left(e^{j\omega}\right)^*.$$

Let $\mathscr{G}$ be a finite-dimensional LTI system with transfer matrix G. Then $y = \mathscr{G}w$ is a WSS stochastic signal with PSD

$$S_y\left(e^{j\omega}\right) = G\left(e^{j\omega}\right) S_w\left(e^{j\omega}\right) G\left(e^{j\omega}\right)^*,$$

and its variance is $\mathbb{E}[|y_t|_2^2] = \|GW\|_2^2$, the square of the H_2-norm of $G(z)W(z)$. In particular, if w is a white noise with (constant) PSD matrix $\sigma^2 I$, for $\sigma \in \mathbb{R}$, then $\mathbb{E}[|y_t|_2^2] = \sigma^2 \|G\|_2^2$.

Proof The formula for the PSD of the signal is discussed in Appendix A. Let $R_t = \mathbb{E}[y_{\tau+t} y_\tau]$ be the autocorrelation matrix of y. We then have

$$E[|y_t|_2^2] = \mathrm{Tr}(R_0) = \mathrm{Tr}\left(\frac{1}{2\pi}\int_{-\pi}^{\pi} S_y\left(e^{j\omega}\right) d\omega\right)$$

$$= \frac{1}{2\pi}\int_{-\pi}^{\pi} \mathrm{Tr}(G\left(e^{j\omega}\right) W\left(e^{j\omega}\right) W\left(e^{j\omega}\right)^* G\left(e^{j\omega}\right)^*) = \|GW\|_2^2.$$

The second interpretation of the H_2 norm concerns impulsive inputs.

Theorem 2.6 *Let $\mathscr{G}$ be a finite-dimensional LTI system with transfer matrix G, with m inputs. Let δ_{t_i} be a discrete impulse signal at time t_i, e_i be the ith standard basis vector in $\mathbb{R}^m$, and $u_i = \delta_{t_i} e_i$ denote an impulse signal on input channel i, for $1 \leq i \leq m$. Then the squared H_2 norm of G represents the sum of the output energies $\sum_{i=1}^{m} \|\mathscr{G}u_i\|_2^2 = \|G\|_2^2$.*

Proof By definition of the impulse response $\{G_t\}_{t \geq 0}$ of the system $\mathscr{G}$, we have

$$\|\mathscr{G}u_i\|_2^2 = \sum_{t=0}^{\infty} |[G_t]_i|_2^2,$$

where $[G_t]_i$ denotes the column i of G_t. It is then immediate that

$$\sum_{i=1}^{m} \|\mathscr{G} u_i\|_2^2 = \sum_{t=0}^{\infty} \sum_{i=1}^{m} \|[G_t]_i\|_2^2 = \sum_{t=0}^{\infty} \|G_t\|_F^2 = \|G\|_2^2.$$

2.3.3.2 Output Perturbation to Process Individual Signals

First, consider a scenario where one wants to process individual signals, while guaranteeing differential privacy for the the adjacency relation (2.3). Suppose that the filter of interest $\mathscr{G}$ is separable (but not necessarily linear), i.e., takes the form

$$\mathscr{G} u = \sum_{i=1}^{n} \mathscr{G}_i u_i. \tag{2.10}$$

Corollary 2.4 *Let $\varepsilon > 0, 1 > \delta > 0$. Let $\mathscr{G}$ be a system defined as in (2.10), with d outputs, and consider the adjacency relation (2.3) for some p-norm. Then the mechanism $M u = \mathscr{G} u + w$, where w is a white noise signal with $w_t \sim Lap(B/\varepsilon)^d$ for $B \geq \rho \max_{1 \leq i \leq n}\{\gamma_{p,1}(\mathscr{G}_i)\}$, is ε-differentially private. If instead $w_t \sim \mathcal{N}(0, \sigma^2 I_d)$, the mechanism is (ε, δ)-differentially private for $\sigma \geq \kappa_{\delta,\varepsilon} \rho \max_{1 \leq i \leq n}\{\gamma_{p,2}(\mathscr{G}_i)\}$.*

In particular, in the last part of Corollary 2.4, if each system $\mathscr{G}_i$ is linear with transfer matrix G_i, and $p_i = 2$ for each i in (2.3), we have $\gamma_{2,2}(\mathscr{G}_i) = \|G_i\|_\infty$.

2.3.3.3 Output Perturbation to Process Collective Signals

Let us now consider collective input signals and the adjacency relations (2.4), (2.5). We can compute exactly the ℓ_2-sensitivity of single input LTI systems, and upper bound it for multiple input systems. This immediately leads to Gaussian mechanisms by Theorem 2.3. For easier reference, we indicate the number of inputs m and outputs q of a system $\mathscr{G}$ as superscripts when we denote its sensitivity $\Delta_p^{m,q}\mathscr{G}$.

Proposition 2.1 (Sensitivity of single input LTI system) *Let $\mathscr{G}$ be an LTI system with transfer function G, one scalar input, q outputs and such that $\|G\|_2 < \infty$. For the adjacency relation (2.4), we have $\Delta_2^{1,q}\mathscr{G} = \rho\|G\|_2$.*

Proof For u and u' adjacent, we have $u - u' = z_0 \delta^{t_0}$ for $|z_0| \leq \rho$, and so

$$\|\mathscr{G}(u - u')\|_2^2 = |z_0|^2 \|\mathscr{G}\delta^{t_0}\|_2^2 \leq \rho^2 \|G\|_2^2,$$

and the bound is attained if $|\alpha_0| = \rho$.   □

For an LTI system $\mathscr{G}$ with multiple inputs, the special case where $\mathscr{G}$ is diagonal, i.e., its transfer matrix is $G(z) = \text{diag}(G_{11}(z), \ldots, G_{mm}(z))$, also leads to a simple sensitivity result. Note that in this case, we have $\|G\|_2^2 = \sum_{i=1}^{m} \|G_{ii}\|_2^2$.

Theorem 2.7 (Diagonal LTI system) *Let $\mathscr{G}$ be a* diagonal *LTI system with m inputs and outputs and transfer function G such that $\|G\|_2 < \infty$. For the adjacency relation (2.5), denoting $R = diag(\rho_1, \ldots, \rho_m)$, we have*

$$\Delta_2^{m,m}\mathscr{G} = \|GR\|_2 = \left(\sum_{i=1}^{m} \|\rho_i G_{ii}\|_2^2\right)^{1/2}.$$

Proof If $\mathscr{G}$ is diagonal, then for u and u' adjacent, we have from (2.8)

$$\|\mathscr{G}(u - u')\|_2^2 = \left\|\sum_{i=1}^{m} z_i \mathscr{G}\delta^{t_i} e_i\right\|_2^2$$
$$= \|\mathrm{col}(z_1 g_{11} * \delta^{t_1}, \ldots, z_m g_{mm} * \delta^{t_m})\|_2^2,$$

where g_{ii} denotes the impulse response of $\mathscr{G}_{ii}$ and $*$ the convolution. Hence,

$$\|\mathscr{G}(u - u')\|_2^2 = \sum_{i=1}^{m} \|z_i g_{ii} * \delta^{t_i}\|_2^2$$
$$= \sum_{i=1}^{m} |z_i|^2 \|G_{ii}\|_2^2,$$

and $|z_i| \leq \rho_i$, for all i. Again the bound is attained if $|z_i| = \rho_i$ for all i.

For multiple input systems, the sensitivity calculations are no longer so straightforward, because the impulses on the various input channels, obtained from the difference of two adjacent signals u, u', all possibly influence any given output. Still, the following result provides simple bounds on the sensitivity.

Theorem 2.8 (Sensitivity bounds for multiple input LTI systems) *Let $\mathscr{G}$ be an LTI system with m inputs, q outputs and such that its transfer function G satisfies $\|G\|_2 < \infty$. For the adjacency relation (2.5), denoting $R = diag(\rho_1, \ldots, \rho_m)$ and $|\rho|_2 = \left(\sum_{i=1}^{m} \rho_i^2\right)^{1/2}$, we have*

$$\|GR\|_2 \leq \Delta_2^{m,q}\mathscr{G} \leq |\rho|_2 \|G\|_2. \tag{2.11}$$

Proof We have $\mathscr{G}(u - u') = \sum_{i=1}^{m} z_i \mathscr{G}\delta^{t_i} e_i$, and moreover $\|G\|_2^2 = \sum_{i=1}^{m} \|\mathscr{G}\delta^{t_i} e_i\|_2^2$ by definition. For the upper bound, we can write

$$\|\mathscr{G}(u - u')\|_2 = \left\|\sum_{i=1}^{m} z_i \mathscr{G}\delta^{t_i} e_i\right\|_2 \leq \sum_{i=1}^{m} |z_i| \|\mathscr{G}\delta^{t_i} e_i\|_2$$
$$\leq |\rho|_2 \left(\sum_{i=1}^{m} \|\mathscr{G}\delta^{t_i} e_i\|_2^2\right)^{1/2},$$

where the last inequality results from the Cauchy–Schwarz inequality.

For the lower bound, let us first take $u' \equiv 0$. Then, consider an adjacent signal u with a single discrete impulse of height ρ_i at time t_i on each input channel i, for $i = 1, \ldots, m$, with $t_1 < t_2 < \cdots < t_m$. Let $\eta > 0$. Denote the "columns" of $\mathscr{G}$ as $\mathscr{G}_i$ for $i = 1, \ldots, m$, i.e., $\mathscr{G}u = \sum_{i=1}^{m} \mathscr{G}_i u_i$. Since $\|G\|_2 < \infty$, $\|G_i u_i\|_2 < \infty$, and hence $|(G_i u_i)_t| \to 0$ as $t \to \infty$. Hence by taking $t_{i+1} - t_i$ large enough for each $1 \leq i \leq m - 1$, i.e., waiting for the effect of impulse i on the output to be sufficiently small, we can choose the signal u such that

$$\|\mathscr{G}u\|_2^2 = \left\| \sum_{i=1}^{m} \mathscr{G}_i u_i \right\|_2^2 \geq \sum_{i=1}^{m} \rho_i^2 \|\mathscr{G}\delta^{t_i} e_i\|_2^2 - \eta.$$

Since this is true for any $\eta > 0$ and $\|\mathscr{G}\delta^{t_i} e_i\|_2^2 = \|G_i\|_2^2$, we get $(\Delta_2^{m,p} \mathscr{G})^2 \geq \|GR\|_2^2 = \sum_{i=1}^{m} \rho_i^2 \|G_i\|_2^2$.

Note that if $\rho_1 = \cdots = \rho_m$, the upper bound on the sensitivity is $\rho_1 \|G\|_2 \sqrt{m}$. We can compare this bound to the situation where G is diagonal, in which case the sensitivity is exactly $\rho_1 \|G\|_2$ from Theorem 2.7. The following example shows that the upper bound of Theorem 2.8 cannot be improved in the general case.

Example 2.1 Consider the system $G(z) = [G_{11}(z), \ldots, G_{1m}(z)]$ with multiple inputs and a single output, with $g_{1i} = \delta_{\tau_i}$ the impulse response of G_{1i}, for some times $\tau_1, \ldots, \tau_m$. Then $\|G\|_2^2 = m$. Now let $u' \equiv 0$ and $u = \sum_{i=1}^{m} \delta^{t_i} e_i$, so that u and u' are adjacent, with $\rho_1 = \cdots = \rho_m = 1$, and moreover let us choose the times t_i such that $\tau_i + t_i$ is a constant, i.e., take $t_i = \kappa - \tau_i$ for some $\kappa \geq \max_i\{\tau_i\}$. Then $\mathscr{G}u = \sum_{i=1}^{m} g_{1i} * u_i = m\delta_\kappa$, and so $\|\mathscr{G}u\|_2^2 = m^2$. This shows that the upper bound of Theorem 2.8 is tight in this case. Note that this happens because all the events of the signal u influence the output at the same time. Indeed, if the times $\tau_i + k_i$ are all distinct, then we get $\|\mathscr{G}u\|_2^2 = m$.

Exact Sensitivity Calculation for Multiple Input Systems

For completeness, we give in this subsection an exact expression for the sensitivity of general finite-dimensional multiple input LTI filters. Let $\mathscr{G}$ be a stable finite-dimensional LTI system with m inputs, q outputs and state space representation

$$\begin{aligned} x_{t+1} &= A x_t + B u_t \\ y_t &= C x_t + D u_t, \end{aligned} \qquad (2.12)$$

with $x_0 = 0$. Recall the definition of the observability Gramian P_0, which is the unique positive semi-definite solution of the equation

$$A^T P_0 A - P_0 + C^T C = 0.$$

Let B_i, D_i be the ith column of the matrix B and D respectively, for $i = 1, \ldots, m$. Finally, define for $i, j \in \{1, \ldots, m\}, i \neq j$, and τ in $\mathbb{Z}$

$$S_{ij}^\tau = \begin{cases} B_i^T (A^{\tau-1})^T C^T D_j + B_i^T (A^\tau)^T P_0 B_j, & \text{if } \tau > 0 \\ D_i^T D_j + B_i^T P_0 B_j, & \text{if } \tau = 0 \qquad (2.13) \\ D_i^T C A^{|\tau|-1} B_j + B_i^T P_0 A^{|\tau|} B_j, & \text{if } \tau < 0. \end{cases}$$

Theorem 2.9 *Let $\mathscr{G}$ be a stable finite-dimensional LTI system with m inputs, q outputs transfer function G and state space representation (2.12). Then, for the adjacency relation (2.5), we have*

$$(\Delta_2^{m,q} \mathscr{G})^2 = \|GR\|_2^2 + \sum_{\substack{i,j=1 \\ i \neq j}}^m \rho_i \rho_j \left(\sup_{\tau \in \mathbb{Z}} |S_{ij}^\tau| \right). \qquad (2.14)$$

Proof In view of (2.8), we have

$$\Delta_2^{m,q} \mathscr{G} = \sup_{|z_i| \leq \rho_i, t_i \geq 0} \left\| \sum_{i=1}^m z_i \mathscr{G} \delta^{t_i} e_i \right\|_2.$$

For $y_i = G\delta^{t_i} e_i$ and $y = \sum_{i=1}^m z_i y_i$, we have

$$\|y\|_2^2 = \sum_{t=0}^\infty \left| \sum_{i=1}^m z_i y_{i,t} \right|^2$$

$$= \sum_{t=0}^\infty \sum_{i=1}^m z_i^2 |y_{i,t}|^2 + \sum_{t=0}^\infty \sum_{\substack{i,j=1 \\ i \neq j}}^m z_i z_j y_{i,t}^T y_{j,t}$$

$$\leq \|GR\|_2^2 + \sum_{\substack{i,j=1 \\ i \neq j}}^m \rho_i \rho_j \left| \sum_{t=0}^\infty y_{i,t}^T y_{j,t} \right|,$$

where $R = \text{diag}(\rho_1, \ldots, \rho_m)$ and the bound can be attained by taking $z_i \in \{-\rho_i, \rho_i\}$, depending on the sign of $S_{ij}^{t_i,t_j} := \sum_{t=0}^\infty y_{i,t}^T y_{j,t}$.

Next, we derive the more explicit expression for the second term $S_{ij}^{t_i,t_j}$, given in the theorem. First, let

$$y_{i,t} = \begin{cases} 0, & t < t_i, \\ D_i, & t = t_i \\ C A^{t-t_i-1} B_i, & t > t_i. \end{cases}$$

Then if $t_i = t_j$, we find that

$$S_{ij}^{t_i,t_j} = D_i^T D_j + B_i^T P_0 B_j,$$

with $P_0 = \sum_{t=0}^\infty (A^t)^T C^T C A^t$ the observability Gramian. If $t_i < t_j$, then

$$S_{ij}^{t_i, t_j} = B_i^T (A^{t_j - t_i - 1})^T C^T D_j + B_i^T (A^{t_j - t_i})^T P_0 B_j.$$

The case $t_i > t_j$ is symmetric. Hence we find that $S_{ij}^{t_i, t_j}$ depends only on the difference $\tau = t_i - t_j$, and our notation (2.13) corresponds to $S_{ij}^{\tau} := S_{ij}^{t_i, t_i + \tau}$.

In (2.14), the maximization over inter-event times τ still needs to be performed and depends on the parameters of the specific system $\mathscr{G}$. This result could be used to evaluate carefully the amount of noise necessary in an output perturbation mechanism. However, it seems too unwieldy to be used in more advanced mechanisms, such as the ones discussed in the next chapters. Still, the expression (2.14) provides some intuition about the way the system dynamics influence its sensitivity. In particular, the second term in (2.14) provides insight into the gap between the sensitivity and the lower bound in (2.11). Note from the expression of S_{ij}^{τ} in (2.13) that one way to decrease the sensitivity of G is to increase sufficiently the time $|t_i - t_j|$ between the events contributed by a single user, in order for $\|A^{|t_i - t_j|}\|_2$ to be small enough. Hence, a lower bound on inter-event times in different streams could be introduced in the adjacency relation to reduce a system's sensitivity. This would weaken the differential privacy guarantee but help in the design of mechanisms with better performance.

We close this section with a result summarizing the output perturbation Gaussian mechanism for collective signals. We state the result for single input systems for simplicity, and at the same time we consider the following generalization of the adjacency condition (2.4). Fix a set $\mathscr{L}$ of LTI systems.

$$\text{Adj}(u, u') \text{ iff } \exists t_0 \in \mathbb{N} \text{ and some system } \mathscr{F} \in \mathscr{L} \text{ s.t. } u_t - u'_t = \begin{cases} 0, & t < t_0, \\ (\mathscr{F}\delta^{t_0})_t, & t \geq t_0. \end{cases} \tag{2.15}$$

Note that in (2.15), $\mathscr{F}\delta^{t_0}$ denotes a signal obtained as the impulse response of $\mathscr{F}$, shifted in time by some t_0, and moreover (2.4) is obtained from (2.15) by taking $\mathscr{L} = \{\alpha \operatorname{Id}, |\alpha| \leq \rho\}$, with Id the identity operator.

Corollary 2.5 *Let $\mathscr{G}$ be an LTI system with one scalar input, q outputs and transfer matrix G. Consider the adjacency relation (2.15) for a given set $\mathscr{L}$. Let $\varepsilon, \delta > 0$. Then the mechanism $Mu = \mathscr{G}u + w$ is (ε, δ)-differentially private when w is a white Gaussian noise with $w_t \sim \mathscr{N}(0, \sigma^2 I)$ for $\sigma \geq \kappa(\delta, \varepsilon) \sup_{F \in \mathscr{L}} \|GF\|_2$.*

2.3.4 Comparison of Input and Output Perturbation Mechanisms

Output perturbation schemes can lead to better performance than input perturbation in certain situations, but only if one can either compute the global sensitivity, or at least provide a tight upper bound for it. For example, the results of Corollary 2.4

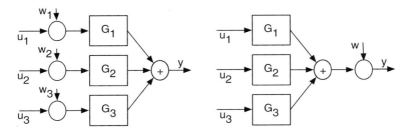

Fig. 2.2 Input (left) versus output (right) perturbation mechanisms for a system of the form (2.10) processing individual signals. The signals w are privacy-preserving noise signals. ©[2014] IEEE. Reprinted, with permission, from Le Ny and Pappas (2014)

show that when processing individual signals, output perturbation schemes add a level of noise that is independent of the number of input signals (i.e., participants), as long as we have a uniform bound on the gains of the subsystems $\mathscr{G}_i$. This is the case for example if all subsystems $\mathscr{G}_i$ are identical. In contrast, the input perturbation mechanism would add noise on each input channel, which would then propagate through the subsystems $\mathscr{G}_i$ and then add at the output. The two types of mechanisms are represented on Fig. 2.2.

The previous remark can be illustrated concretely through the following example. Consider a system of the form (2.10) and suppose that all systems $\mathscr{G}_i$ are the same moving average filter with window size l, i.e., $(\mathscr{G}_i u_i)_t = \frac{1}{l} \sum_{k=0}^{l-1} u_{i,t-k}$, for all $1 \le i \le n$. Then one can compute $\|G_i\|_\infty = 1$ and $\|G_i\|_2 = 1/\sqrt{l}$. The adjacency relation is (2.3) with $\rho = 1$ and $p = 2$. The Gaussian input perturbation mechanism produces an output $y_1 = \sum_{i=1}^{n} \mathscr{G}_i u_i + w_1$, with the noise signal $w_1 := \sum_{i=1}^{n} \mathscr{G}_i w_i$ having a steady-state variance of $\kappa_{\delta,\varepsilon}^2 \sum_{i=1}^{n} \|G_i\|_2^2 = \kappa_{\delta,\varepsilon}^2 n/l$, as recalled in Theorem 2.5. In contrast, the output perturbation mechanism produces $y_2 = \sum_{i=1}^{n} \mathscr{G}_i u_i + w_2$, where w_2 is a white Gaussian noise with variance $\kappa_{\delta,\varepsilon}^2$. Hence, when performance is measured by the mean-squared error (MSE), the output mechanism is preferable to the input mechanism as soon as the number of participants is greater than the length l of the averaging window.

References

Bolot et al J (2013) Private decayed predicate sums on streams. In: Proceedings of the 16th international conference on database theory, Genoa, Italy, pp 284–295

Chan T-HH, Shi E, Song D (2011) Private and continual release of statistics. ACM Trans Inf Syst Secur 14(3):26:1–26:24

Dwork C, Roth A (2014) The algorithmic foundations of differential privacy. Found Trends Theor Comput Sci 9(3–4):211–407

Dwork C et al (2006a) Calibrating noise to sensitivity in private data analysis. In: Proceedings of the third theory of cryptography conference, New York, NY, pp 265–284

Dwork C et al (2006b) Our data, ourselves: privacy via distributed noise generation. In: Proceedings of the 24th annual international conference on the theory and applications of cryptographic techniques (EUROCRYPT), St. Petersburg, Russia, pp 486–503

Dwork C et al (2010) Differential privacy under continual observations. In: Proceedings of the ACM symposium on the theory of computing (STOC), Cambridge, MA

Hart GW (1992) Nonintrusive appliance load monitoring. Proc IEEE 80(12):1870–1891

Le Ny J, Mohammady M (2018) Differentially private MIMO filtering for event streams. IEEE Trans Autom Control 63(1)

Le Ny J, Pappas GJ (2014) Differentially private filtering. IEEE Trans Autom Control 59(2):341–354

Molina-Markham A et al (2010) Private memoirs of a smart meter. In: Proceedings of the 2nd ACM workshop on embedded sensing systems for energy-efficiency in building, New York, NY, USA, pp 61–66

Pyrgelis A, Troncoso C, De Cristofaro E (2017) What does the crowd say about you? evaluating aggregation-based location privacy. In: Proceedings of privacy enhancing technologies

Xu F et al (2017) Trajectory recovery from ash: user privacy is not preserved in aggregated mobility data. In: Proceedings of the 26th international conference on world wide web, pp 1241–1250

Chapter 3
A Two-Stage Architecture for Differentially Private Filtering

3.1 Introduction

[1]In Chap. 2 we discussed two basic noise-additive architectures to release differentially private numerical signals, namely, using input perturbation or output perturbation. This chapter presents a more general two-stage architecture to enforce differential privacy, consisting of signal pre-processing and post-processing stages placed around the additive noise mechanism. Designs based on this architecture can often achieve significant performance improvements compared to input and output perturbation, for the same level of privacy. A general design methodology for two-stage mechanisms is presented. Its application and potential performance gains are illustrated by introducing the differentially private zero-forcing equalization mechanism for the processing of collective signals, as defined in Chap. 2.

3.2 Design Methodology for Two-Stage Mechanisms

A general goal in privacy-preserving signal processing is to design an approximation of a given real-time signal processing system that provides a differential privacy guarantee for the input signals from which published output signals are computed. The quality of the approximation is measured by how close the outputs produced by the privacy-preserving system are to the signals we would obtain in the absence of privacy constraint, see Fig. 3.1. Performance measures such as the mean-squared error (MSE) between the private and non-private signals provide quantitative means to evaluate and optimize a given design.

---

[1]Some of the text in Sects. 3.4.1, 3.4.2 and 3.5 of this chapter is reprinted, with permission, from Le Ny and Mohammady (2018) (©[2018] IEEE).

© The Author(s), under exclusive license to Springer Nature Switzerland AG 2020
J. Le Ny, *Differential Privacy for Dynamic Data*,
SpringerBriefs in Control, Automation and Robotics,
https://doi.org/10.1007/978-3-030-41039-1_3

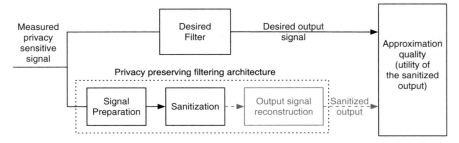

Fig. 3.1 Architecture for privacy preserving filtering. Signals represented by dashed lines (after sanitization) are differentially private. ©[2016] IEEE. Reprinted, with permission, from Cortés et al. (2016)

Figure 3.1 presents a signal processing architecture generalizing the input and output perturbation architectures presented in Chap. 2, which turns out to be convenient to optimize the quality of a differentially private approximation. One motivation for this architecture is that leaving raw white noise on the published signal with an output perturbation scheme appears intuitively to be suboptimal, as this noise should be somehow further filtered to improve the quality of the output. The architecture of Fig. 3.1 fixes this issue and can be interpreted as follows. A given privacy-sensitive input signal is first appropriately "shaped" by an input block playing the role of an intelligent sensor or transmitter. The resulting signal is then sanitized through an additive noise mechanism, which can be viewed as a noisy transmission channel. This step provides a differential privacy guarantee, e.g., through the addition of Laplace or Gaussian noise according to Theorem 2.3. Finally, a terminal block can be viewed as a communication receiver and attempts to remove the effect of the input block and sanitization process to estimate the desired output signal as accurately as possible. Note that the design of the terminal block depends on the input block, which is known. This architecture enforces differential privacy as long as the sanitization mechanism does and the terminal block does not re-access the original privacy-sensitive signal, by the resilience to post-processing property mentioned after Theorem 1.2. We recover the input and output perturbation mechanisms if the signal preparation or output signal reconstruction block are taken as the identity operator respectively, and the remaining block is the desired filter.

Designing and optimizing the two-stage architecture of Fig. 3.1 can be done as follows:

1. First, we choose the adjacency relation on the space of measured signals and the sanitization mechanism to provide differential privacy. For example, the Gaussian mechanism will add Gaussian noise proportional to the sensitivity of the signal preparation block, which itself depends on the choice of adjacency relation on the input signals. We also choose a measure of performance to evaluate the approximation quality between the desired and privacy-preserving signals.

2. Second, we choose a structure for the output signal reconstruction block. It is in general motivated by the choice of performance measure and possibly by the availability of any public information one might have about the input signals, which could help in the estimation of the desired output signal from the noisy output of the sanitization step. Since the signal preparation block has not yet been specified at this stage, we express the output signal reconstruction block as a *function of the signal preparation block*.
3. Finally, we express the performance as a function of the signal preparation block only, and choose the best such block optimizing the performance.

In the rest of this chapter, we illustrate the application of this methodology to the differentially private processing of collective signals, as defined in Sect. 2.3.1. In the absence of public information about the statistics of these input signals, one possible choice for the output signal reconstruction block is to simply invert the effect of the signal preparation block, a process called zero-forcing equalization (ZFE) in the communication literature. We therefore call the resulting mechanism "zero-forcing equalization mechanism". To close this section, it is also interesting to note the similarity of the design problem for the architecture of Fig. 3.1 and classical joint transmitter-receiver optimization problems (Salz 1985; Yang and Roy 1994). As previously noted, the input block plays the role of a transmitter and the output block that of a receiver. This type of architecture has also been used for example in Li and Miklau (2012) to provide differentially private answers to linear queries about static databases, in Sankar et al. (2010) for an information theoretic definition of privacy, or in Tanaka et al. (2017) to study a sensor design problem for communication-constrained estimation.

3.3 Two-Block Approximation Architecture for LTI Filters

Consider a scenario where m sensors each report at regular intervals a scalar value $u_{i,t} \in \mathbb{R}$, for $t \geq 0$, $i = 1, \ldots, m$. These m scalar signals correspond to privacy-sensitive collective signals as described in Sect. 2.3.1. We consider the adjacency relation (2.4) when $m = 1$ and its generalizaton (2.5) when $m > 1$. Let $u_t = [u_{1,t}, \ldots, u_{m,t}]^T \in \mathbb{R}^m$. We would like to publish in real-time a signal $y = Fu$ computed from the signal u, with $y_t \in \mathbb{R}^p$, where the filter F, which is given and depends on the application of interest, is assumed to be a causal LTI system with m inputs and p outputs. The privacy-sensitive nature of the signals u prevents us from releasing exactly the signal Fu, since F could be a partially or completely invertible system for example. Instead, we release a sanitized, differentially private approximation $\hat{y}$ of y and evaluate the quality of this approximation by a measure such as the steady-state average MSE $\lim_{T \to \infty} \frac{1}{T} \sum_{t=0}^{T-1} \mathbb{E}[|y_t - \hat{y}_t|_2^2]$, which we want to minimize.

The approximation architecture shown on Fig. 3.2 is an instantiation of the general architecture of Fig. 3.1. We need to design the filters G and H, which we assume here to be LTI systems. The privacy-sensitive input signal u is first processed by a

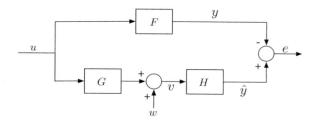

filter G. Then, according to the general Theorem 2.3, the signal $v = Gu + w$ can be made (ε, δ)-differentially private by adding white Gaussian noise w with variance $(\kappa_{\delta,\varepsilon} \, \Delta_2 G)^2 I_{p'}$, where p' is the chosen number of outputs of G and $\Delta_2 G$ is the sensitivity of G, whose computation is discussed in Sect. 2.3.3.3. By resilience to post-processing, the signal $\hat{y}$ computed from v to approximate y is differentially private no matter what the system H is, see Sect. 1.3.3. Again, note that the input perturbation mechanism (see Corollary 2.2) is recovered for $G = \mathrm{Id}$ and $H = F$, and the output perturbation mechanism (see Sect. 2.3.3.3) for $G = F$ and $H = \mathrm{Id}$.

We can try to construct G and H to minimize the MSE between $y = Fu$ and $\hat{y} = HGu + Hw$

$$e_{mse}(G, H) = \lim_{T \to \infty} \frac{1}{T} \sum_{t=0}^{T-1} \mathbb{E}\left[|y_t - \hat{y}_t|_2^2\right] \tag{3.1}$$

$$= \lim_{T \to \infty} \frac{1}{T} \sum_{t=0}^{T-1} \mathbb{E}\left[|((F - HG)u)_t - (Hw)_t|_2^2\right].$$

At this point, we have carried out Step 1 of the design procedure outlined in the Sect. 3.2. We chose the Gaussian mechanism for sanitization and the MSE as performance measure. Note that the Gaussian noise could possibly be replaced by Laplace noise for ε-differential privacy, but the computation of the ℓ_1-sensitivity of linear systems for the adjacency relation (2.4) becomes more difficult. Hence, in the following, we restrict our discussion to Gaussian noise. Step 2 of the procedure is carried out in the next section, starting with a design choice for the block H.

3.4   Zero-Forcing Equalization Mechanism

For the ZFE mechanism, the output block H simply inverts G and applies the desired filter F to obtain the approximation signal $\hat{y}$. In other words, we fix the system H to be of the form $H = FL$, with L a causal left inverse of the pre-filter G, i.e., such that $LG = \mathrm{Id}$, assuming such an L exists. Referring again to the notation on Fig. 3.2, we have now

$$\hat{y} = Hz = FLGu + Hw = Fu + Hw = y + Hw,$$

and so the approximation error between y and $\hat{y}$ is directly measured by the level of noise Hw. Because w is white Gaussian noise and H is time-invariant, the error signal $e := y - \hat{y} = -Hw$ is in this case *stationary* (even if u is not), Gaussian, zero-mean, and its variance is that of w multiplied by $\|H\|_2^2$, following Theorem 2.5. The MSE (3.1) between y and $\hat{y}$ can now be expressed explicitly as

$$e_{mse}^{ZFE}(G, L) = \lim_{T \to \infty} \frac{1}{T} \sum_{t=0}^{T-1} \mathbb{E}[y_t - \hat{y}_t|_2^2] = \lim_{T \to \infty} \frac{1}{T} \sum_{t=0}^{T-1} \mathbb{E}\left[|(Hw)_t|_2^2\right] \quad (3.2)$$

$$= \mathbb{E}[|y_t - \hat{y}_t|^2] = \frac{1}{2\pi} \int_{-\pi}^{\pi} \text{Tr}\left(S_e\left(e^{j\omega}\right)\right) d\omega \quad (3.3)$$

$$= (\kappa_{\delta,\varepsilon}\Delta_2 G)^2 \|H\|_2^2 = (\kappa_{\delta,\varepsilon}\Delta_2 G)^2 \|FL\|_2^2, \quad (3.4)$$

with S_e the PSD of the error signal e and the constraint $LG = \text{Id}$. Note that (3.3) expresses the fact that the error signal variance for this mechanism is independent of time, a consequence of the fact that the error signal is stationary.

The expression (3.4) for the ZFE mechanism includes again as special cases the input perturbation mechanism ($G = L = \text{Id}$, $H = F$) and output perturbation mechanism ($G = F$, $H = \text{Id}$, even if F does not have a left inverse). Suppose F has a single input, i.e., $m = 1$. Then by Theorem 2.1 we have $\Delta_2 F = \rho \|F\|_2$, and then we see from (3.4) that both the input and output perturbation mechanism with Gaussian noise give the same MSE, equal to $\rho^2 \kappa_{\delta,\varepsilon}^2 \|F\|_2^2$.[2] The noises seen at the output (or error signals) have different frequency content however (different PSD), since in the input perturbation case the Gaussian noise passes through F (hence the PSD of the error at the output is $\rho^2 \kappa_{\delta,\varepsilon}^2 F(e^{i\omega})F(e^{i\omega})^*$ by Theorem 2.5), whereas for output perturbation the spectral content of the output noise is flat (PSD equal to $\rho^2 \kappa_{\delta,\varepsilon}^2 \|F\|_2^2 I$). This motivates the introduction of a frequency-weighted version of the MSE performance measure (3.2), as follows

$$e_{mse,T}^{ZFE}(G, L) = \frac{1}{2\pi} \int_{-\pi}^{\pi} \text{Tr}(T(e^{j\omega})S_e(e^{j\omega})T(e^{j\omega})^*) d\omega, \quad (3.5)$$

where T is a frequency weighting function (with $T(e^{j\omega}) \in \mathbb{C}^{p \times p}$ for each ω), placing a larger weight on frequencies in the error signal e that we wish to reduce more. The expression (3.2) corresponds to $T \equiv 1$. Given that $e = -FLw$, by Theorem 2.5 the PSD S_e of the error signal is

---

[2]Note the difference with the discussion comparing input and output perturbation for the processing of individual signals in Sect. 2.3.4, where one scheme could be better than the other. This conclusion also does not necessarily hold for the Laplace mechanism. This shows that the choice of input signal space, adjacency relation, etc., has an important impact on the conclusions one can draw about particular differentially private mechanisms.

$$S_e(e^{j\omega}) = F(e^{j\omega})L(e^{j\omega})L(e^{j\omega})^*F(e^{j\omega})^* \times \rho^2\kappa_{\delta,\varepsilon}^2 \|G\|_2^2,$$

and the frequency weighted performance measure (3.5) reduces to the unweighted one by replacing the specified filter F by TF. Hence, it is sufficient to consider in the following discussion the unweighted performance measure, equivalently $T \equiv 1$, as we do to simplify the notation.

In the rest of this section, we show how to choose an optimal LTI system G to minimize (3.2) or (3.5), to improve on the input and output perturbation mechanisms. This amounts to carrying out Step 3 of the general design methodology presented in Sect. 3.2. Since the expression (3.4) depends on the computation of the sensitivity $\Delta_2 G$, which by Sect. 2.3.3.3 is only known exactly for single input systems, we divide the discussion between single input and multiple input systems.

3.4.1   Single-Input Case

If F on Fig. 3.2 is a single-input single-output (SISO) or single-input multiple-output (SIMO) filter, i.e., $m = 1$, then so is G (we will soon see that it is in fact sufficient to take G SISO). Recall that we consider the adjacency relation (2.4) on privacy sensitive collective input signals. In this case, Proposition 2.1 says that the sensitivity $\Delta_2 G$ appearing the MSE expression (3.4) is equal to $\rho\|G\|_2$. For the ZFE mechanism, we restrict slightly the design space by requiring that G has a left inverse L, and we choose $H = FL$ on Fig. 3.2. The next theorem gives a condition on the optimal filters G and L minimizing (3.4) for the ZFE mechanism. This condition allows us to construct G and L and shows that one can take these optimal filters to be SISO. Note that if F has p outputs (with possibly $p = 1$), then for each ω, $F(e^{j\omega})$ is a complex-valued p-dimensional vector and $|F(e^{j\omega})|_2$ denotes the Euclidean norm of that vector.

Theorem 3.1 *Let F be a SIMO LTI system with $\|F\|_2 < \infty$. For any SIMO LTI system G such that $\|G\|_2 < \infty$ and with a left-inverse L, we have*

$$e_{mse}^{ZFE}(G, L) \geq \rho^2\kappa_{\delta,\varepsilon}^2 \left(\frac{1}{2\pi}\int_{-\pi}^{\pi}|F(e^{j\omega})|_2 \, d\omega\right)^2. \tag{3.6}$$

Suppose moreover that F satisfies the Paley–Wiener condition $\frac{1}{2\pi}\int_{-\pi}^{\pi}\ln|F(e^{j\omega})|_2$ $d\omega > -\infty$. Then there exists a SISO LTI system G with causal inverse L such that $\|G\|_2 < \infty$,

$$|G(e^{j\omega})|^2 = |F(e^{j\omega})|_2 \text{ for almost every } \omega \in [-\pi, \pi), \tag{3.7}$$

and any such G, L achieves the lower bound (3.6) and satisfies $\|G\|_2 = \|FL\|_2$.

Finding G SISO satisfying (3.7), $\|G\|_2 < \infty$ and with causal inverse $L = G^{-1}$ is a classical spectral factorization problem, see Kailath et al. (2000) for example for more details.

Proof On Fig. 3.2, consider a first stage $G(z) = [G_1(z), \ldots, G_q(z)]^T$ with $\|G\|_2 < \infty$, taking the input signal u and producing q intermediate outputs, for some $q \geq 1$. These outputs are then perturbed by Gaussian noise. A (not necessarily causal) left inverse of G is an LTI filter $L(z) = [L_1(z), \ldots, L_q(z)]$ satisfying $L(z)G(z) = \sum_{i=1}^q L_i(z)G_i(z) = 1$. The second stage is $H = FL$. Let us also introduce the para-hermitian conjugate of L as $M(z) = L((1/z^*))^*$, so that $M(e^{j\omega}) = L(e^{j\omega})^*$. Thus, for all ω,

$$|M(e^{j\omega})|_2 = |L(e^{j\omega})|_2, \tag{3.8}$$

$$\text{and } M(e^{j\omega})^* G(e^{j\omega}) = 1. \tag{3.9}$$

From Proposition 2.1, the ℓ_2-sensitivity of the first stage for input signals that are adjacent according to (2.4) is $\rho \|G\|_2$. Hence, according to Theorem 2.3, adding a white Gaussian noise w to the output of G with covariance matrix $\rho^2 \kappa_{\delta,\varepsilon}^2 \|G\|_2^2 I_q$ is sufficient to ensure that the signal $v = Gu + w$ on Fig. 3.2 is differentially private. The MSE (3.4) for this mechanism now reads

$$e_{mse}^{ZFE}(G, L) = \rho^2 \kappa_{\delta,\varepsilon}^2 \|G\|_2^2 \|FL\|_2^2.$$

We are thus led to consider the minimization of $\|FL\|_2^2 \|G\|_2^2$ over the pre-filters G such that $\|G\|_2 < \infty$ and $\|FL\|_2 < \infty$. We have

$$
\begin{aligned}
&\|FL\|_2^2 \|G\|_2^2 \\
&= \frac{1}{2\pi} \int_{-\pi}^{\pi} \text{Tr}(L^*(e^{j\omega}) F^*(e^{j\omega}) F(e^{j\omega}) L(e^{j\omega})) d\omega \times \frac{1}{2\pi} \int_{-\pi}^{\pi} \text{Tr}(G^*(e^{j\omega}) G(e^{j\omega})) d\omega \\
&= \frac{1}{2\pi} \int_{-\pi}^{\pi} |F(e^{j\omega})|_2^2 |L(e^{j\omega})|_2^2 d\omega \times \frac{1}{2\pi} \int_{-\pi}^{\pi} |G(e^{j\omega})|_2^2 d\omega \\
&= \frac{1}{2\pi} \int_{-\pi}^{\pi} |F(e^{j\omega})|_2^2 |M(e^{j\omega})|_2^2 d\omega \times \frac{1}{2\pi} \int_{-\pi}^{\pi} |G(e^{j\omega})|_2^2 d\omega,
\end{aligned}
$$

where in the last equality we used (3.8).

Now consider the inner product $\langle f, g \rangle = \frac{1}{2\pi} \int_{-\pi}^{\pi} f(e^{j\omega})^* g(e^{j\omega}) d\omega$ on the space of 2π-periodic functions with values in $\mathbb{C}^q$. By the Cauchy–Schwarz inequality for this inner product applied to the functions $\omega \mapsto |F(e^{j\omega})|_2 M(e^{j\omega})$ and $\omega \mapsto G(e^{j\omega})$, we obtain the bound

$$\|FL\|_2^2\|G\|_2^2 \geq \left(\frac{1}{2\pi} \int_{-\pi}^{\pi} |F(e^{j\omega})|_2 M(e^{j\omega})^* G(e^{j\omega}) d\omega \right)^2$$

$$\geq \left(\frac{1}{2\pi} \int_{-\pi}^{\pi} |F(e^{j\omega})|_2 \, d\omega \right)^2, \text{ using (3.9)},$$

which gives (3.6). Moreover, the two sides in the Cauchy–Schwarz inequality are equal, i.e., the bound is attained, if

$$G(e^{j\omega}) = |F(e^{j\omega})|_2 M(e^{j\omega}). \tag{3.10}$$

Note that this condition does not depend on q, hence we can simply take $q = 1$. Since $|F(e^{j\omega})|_2$ is an nonnegative function on the unit circle, integrable on $[-\pi, \pi]$ (consequence of $\|F\|_2 < \infty$), if it satisfies the Paley–Wiener condition, it has indeed a (minimum phase) spectral factor G with causal inverse satisfying (3.7) almost everywhere (Kailath et al. 2000, p. 199), i.e., $G(e^{j\omega})G(e^{j\omega})^* = |F(e^{j\omega})|_2$. Multiplying both side of this equality by $M(e^{j\omega})$ and using (3.9), we obtain (3.10), and so such as SISO filter G attains the lower bound.

Finally, $\|FL\|_2 = \|G\|_2$ is a straightforward consequence of the stronger pointwise condition (3.7).   □

3.4.2   Multiple-Input Case

For multiple input systems, the exact expression of the sensitivity with respect to the adjacency relation (2.5), given in Theorem 2.9, is more complicated than for SIMO systems, because of the superposition at the output of the effect of one individual on the different input channels. As a result, we do not attempt here to use that expression to optimize MIMO ZFE mechanisms, but we introduce further potential conservatism by using only diagonal input blocks G, for which the exact sensitivity is known, see Theorem 2.7. In effect, this reduces the MIMO mechanism design problem to a set of SIMO problems, one for each input channel. In addition, we derive bounds on the achievable performance improvement if more general (i.e., not necessarily diagonal) input blocks were used.

3.4.2.1   Diagonal Pre-filter Optimization

Hence, let us now assume that F has $m > 1$ inputs. We write $F(z) = [F_1(z), \ldots, F_m(z)]$, with F_i a $p \times 1$ vector of transfer functions. As mentioned above, the idea is to restrict our design to pre-filters G that are $m \times m$ and diagonal, with squared sensitivity equal to $(\Delta_2^{m,m} G)^2 = \|GR\|_2^2 = \sum_{i=1}^{m} \|\rho_i G_{ii}\|_2^2$, where $R = \mathrm{diag}(\rho_1, \ldots, \rho_m)$ according to Theorem 2.7. Thus, we consider the architecture shown on Fig. 3.3, where the signal w is a white Gaussian noise with covariance matrix $\kappa_{\delta,\varepsilon}^2 \|GR\|_2^2 I_m$,

Fig. 3.3 (Suboptimal) ZFE mechanism for a MIMO system $Fu = \sum_{i=1}^{m} F_i u_i$, and a diagonal pre-filter $G(z) = \text{diag}(G_{11}(z), \ldots, G_{mm}(z))$. Here $F_i(z)$ is a $p \times 1$ transfer matrix, for $i = 1, \ldots, m$. ©[2018] IEEE. Reprinted, with permission, from Le Ny and Mohammady (2018)

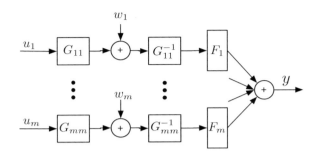

making the signals $G_{ii}u_i + w_i$, $i = 1, \ldots, m$, differentially private. The problem of optimizing the pre-filters G_{ii} can then be seen as designing m SIMO mechanisms.

Theorem 3.2 *Let $F = [F_1, \ldots, F_m]$ be a MIMO LTI system with $\|F\|_2 < \infty$. For any diagonal LTI system $G(z) = \text{diag}(G_{11}(z), \ldots, G_{mm}(z))$ such that $\|G\|_2 < \infty$, with inverse L, we have*

$$e_{mse}^{ZFE}(G, L) \geq \kappa_{\delta,\varepsilon}^2 \left(\frac{1}{2\pi} \int_{-\pi}^{\pi} \sum_{i=1}^{m} \rho_i |F_i(e^{j\omega})|_2 \, d\omega \right)^2. \tag{3.11}$$

If moreover each F_i satisfies the Paley–Wiener condition $\frac{1}{2\pi} \int_{-\pi}^{\pi} \ln |F_i(e^{j\omega})|_2 \, d\omega > -\infty$, this lower bound is attained by some SISO systems G_{ii} with causal inverses G_{ii}^{-1}, such that $\|\rho_i G_{ii}\|_2 = \|F_i G_{ii}^{-1}\|_2 < \infty$ and $\rho_i |G_{ii}(e^{j\omega})|^2 = |F_i(e^{j\omega})|_2$, for almost every $\omega \in [-\pi, \pi)$.

In other words, the best diagonal pre-filter for the MIMO ZFE mechanism can be obtained from m spectral factorizations of the functions $\omega \mapsto \frac{1}{\rho_i} |F_i(e^{j\omega})|_2$, $i = 1, \ldots, m$.

Proof Following the same reasoning as in the proof of Theorem 3.1, the MSE for the mechanism of Fig. 3.3 can be expressed as

$$e_{mse}^{ZFE}(G) = \kappa_{\delta,\varepsilon}^2 \|GR\|_2^2 \|FG^{-1}\|_2^2, \tag{3.12}$$

with $G^{-1}(z) = \text{diag}(G_{11}(z)^{-1}, \ldots, G_{mm}(z)^{-1})$, assuming that the inverses exist. Now remark that $\|FG^{-1}\|_2^2 = \frac{1}{2\pi} \int_{-\pi}^{\pi} \sum_{i=1}^{m} \frac{|F_i(e^{j\omega})|_2^2}{|G_{ii}(e^{j\omega})|_2^2} \, d\omega$. Hence from the Cauchy–Schwarz inequality again, we obtain the lower bound

$$e_{mse}^{ZFE}(G) \geq \kappa_{\delta,\varepsilon}^2 \left(\frac{1}{2\pi} \int_{-\pi}^{\pi} \sum_{i=1}^{m} \frac{|F_i(e^{j\omega})|_2}{|G_{ii}(e^{j\omega})|} |\rho_i G_{ii}(e^{j\omega})| d\omega \right)^2$$

$$e_{mse}^{ZFE}(G) \geq \kappa_{\delta,\varepsilon}^2 \left(\frac{1}{2\pi} \int_{-\pi}^{\pi} \sum_{i=1}^{m} \rho_i |F_i(e^{j\omega})|_2 \, d\omega \right)^2,$$

and this bound is attained if $\rho_i |G_{ii}(e^{j\omega})| = \frac{|F_i(e^{j\omega})|_2}{|G_{ii}(e^{j\omega})|}$, i.e., $\rho_i |G_{ii}(e^{j\omega})|^2 = |F_i(e^{j\omega})|_2$, for $i = 1, \ldots, m$.

3.4.2.2 Comparison with Non-diagonal Pre-filters

For F a general MIMO system, it is possible that we could achieve a better performance with a ZFE mechanism where G is not diagonal, i.e., by combining the inputs before adding the privacy-preserving noise. To estimate how much could potentially be gained by carrying out this more involved optimization over general pre-filters G rather than just diagonal pre-filters, the following theorem provides a lower bound on the MSE achievable by *any* ZFE mechanism (still adding white Gaussian noise at the output of G, proportional to the ℓ_2-sensitivity of G).

Theorem 3.3 *Let $F = [F_1, \ldots, F_m]$ be a MIMO LTI system with $\|F\|_2 < \infty$. For any $m \times m$ LTI system G such that $\|G\|_2 < \infty$, with left inverse L such that $\|FL\|_2 < \infty$, we have*

$$e_{mse}^{ZFE}(G, L) \geq \kappa_{\delta,\varepsilon}^2 \left(\frac{1}{2\pi} \int_{-\pi}^{\pi} \|F(e^{j\omega})R\|_* \, d\omega \right)^2, \tag{3.13}$$

where $\|F(e^{j\omega})R\|_*$ denotes the nuclear norm (i.e., sum of singular values) of the matrix $F(e^{j\omega})R$.

The *lower bound* (3.13) on the MSE achievable with a general pre-filter in a ZFE mechanism should be compared to the performance (3.11) that can actually be achieved with diagonal pre-filters. Note that these bounds coïncide for $m = 1$. For $m > 1$, the gap depends on the difference between the integrals of two different norms of $F(e^{j\omega})R$. Note also from (3.11), (3.13) that the dependency of the ZFE mechanism performance with respect to the parameters ε, δ is captured only by the factor $\kappa_{\delta,\varepsilon}$.

Proof With $R = \mathrm{diag}(\rho_1, \ldots, \rho_m)$ as usual, we define $\check{G} = GR$ and $\check{L} = R^{-1}L$, so that $\check{L}\check{G} = I$. Let $\check{F} = FR$. With the lower bound of Theorem 2.8, designing a ZFE mechanism based on sensitivity as above would require adding a noise with variance at least $\kappa_{\delta,\varepsilon}^2 \|\check{G}\|_2^2$. This would lead to an MSE at least equal to $\kappa_{\delta,\varepsilon}^2 \|\check{G}\|_2^2 \|\check{F}\check{L}\|_2^2$. Now note that

$$\|\check{F}\check{L}\|_2^2 = \frac{1}{2\pi} \int_{-\pi}^{\pi} \mathrm{Tr}(\check{F}(e^{j\omega})\check{L}(e^{j\omega})\check{L}(e^{j\omega})^* \check{F}(e^{j\omega})^*) d\omega$$

$$= \frac{1}{2\pi} \int_{-\pi}^{\pi} \mathrm{Tr}(\check{F}(e^{j\omega})^* \check{F}(e^{j\omega})\check{L}(e^{j\omega})\check{L}(e^{j\omega})^*) d\omega$$

$$= \frac{1}{2\pi} \int_{-\pi}^{\pi} \mathrm{Tr}(A(e^{j\omega})^2 \check{L}(e^{j\omega})\check{L}(e^{j\omega})^*) d\omega$$

$$= \frac{1}{2\pi} \int_{-\pi}^{\pi} \mathrm{Tr}(A(e^{j\omega})\check{L}(e^{j\omega})\check{L}(e^{j\omega})^* A(e^{j\omega})) d\omega,$$

where for all ω, $A(e^{j\omega})$ is the unique Hermitian positive-semidefinite square root of $\check{F}(e^{j\omega})^*\check{F}(e^{j\omega})$, i.e., $A(e^{j\omega})^2 = \check{F}(e^{j\omega})^*\check{F}(e^{j\omega})$ (Horn and Johnson 2012, Theorem 7.2.6). Then, once again from the Cauchy–Schwarz inequality, now for the inner product $\langle M, N \rangle = \frac{1}{2\pi}\int_{-\pi}^{\pi}\mathrm{Tr}(M(e^{j\omega})^*N(e^{j\omega}))d\omega$,

$$\|GK\|_2^2\,\|FL\|_2^2 = \|\check{G}\|_2^2\,\|\check{F}\check{L}\|_2^2$$

$$= \frac{1}{4\pi^2}\left[\int_{-\pi}^{\pi}\mathrm{Tr}(\check{G}(e^{j\omega})^*\check{G}(e^{j\omega}))d\omega\right] \times \left[\int_{-\pi}^{\pi}\mathrm{Tr}(A(e^{j\omega})\check{L}(e^{j\omega})\check{L}(e^{j\omega})^*A(e^{j\omega}))d\omega\right]$$

$$\geq \left(\frac{1}{2\pi}\int_{-\pi}^{\pi}\mathrm{Tr}(A(e^{j\omega})\check{L}(e^{j\omega})\check{G}(e^{j\omega}))d\omega\right)^2$$

and so $e_{mse}^{ZFE}(G) \geq \kappa_{\delta,\varepsilon}^2\left(\frac{1}{2\pi}\int_{-\pi}^{\pi}\|F(e^{j\omega})R\|_*d\omega\right)^2$,

where $\|F(e^{j\omega})R\|_* = \mathrm{Tr}(A(e^{j\omega}))$ is the nuclear norm of the matrix $F(e^{j\omega})R$. □

3.5 Application to Privacy-Preserving Estimation of Building Occupancy

As an illustration of the methodology described in this chapter, in this section we use the ZFE mechanism to design a privacy-preserving filter estimating occupancy in an office building equipped with motion detection sensors. Our presentation is brief and the reader is referred to Le Ny and Mohammady (2018) for additional details. We consider the data collected during a sensor network deployment experiment described in Wren et al. (2007). The original dataset contains the traces of 213 sensors placed a few meters apart and spread over two floors of a building, where each sensor recorded with millisecond accuracy over a year the exact times at which it detected some motion. Datasets of such fine spatial and temporal granularity raise clear privacy concerns and Ivanov et al. (2007) for example shows how to re-identify individual trajectories from it. We downsampled the data in space and time, summing all the events recorded by several sufficiently close sensors over 3 min intervals. We then obtained 15 input signals u_i, $i = 1, \ldots, 15$, corresponding to 15 spatial zones (each zone covered by a cluster of about 14 sensors), with a discrete-time period corresponding to 3 min and where $u_{i,t}$ is the number of events detected by all the sensors in zone i during period t. If we assume for example that during a given discrete-time period, a single individual can activate at most 4 sensors in any zone, then $\rho_i = 4$ for $1 \leq i \leq 15$ in the adjacency relation (2.5), which also assumes that a single individual only activates the sensors in a given zone once over the time interval of interest.

Suppose that we want to compute simultaneously and in real-time the following three outputs from the 15 input signals

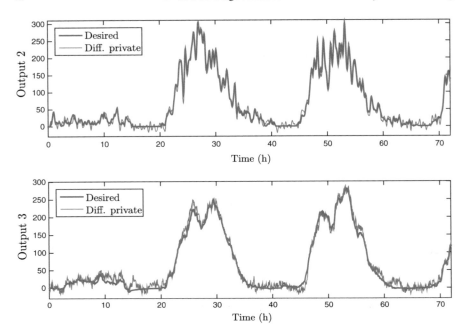

Fig. 3.4 Sample paths over 72 h (the sampling period is 3 min) for the outputs 2 and 3 of our differentially private approximation of filter (3.14), shown together with the desired outputs. The privacy parameters are $\varepsilon = \ln 5$, $\delta = 0.05$, $\rho_i = 4$ for $1 \leq i \leq 15$. The average ℓ_2 norm per period for the error signal corresponding to each figure is 9.3 and 14.5 respectively. ©[2018] IEEE. Reprinted, with permission, from Le Ny and Mohammady (2018)

$$\begin{bmatrix} y_1 \\ y_2 \\ y_3 \end{bmatrix} = \begin{bmatrix} f_1(z)\mathbf{1}_{1\times 5} & \mathbf{0}_{1\times 10} \\ \mathbf{0}_{1\times 4} & f_2(z)\mathbf{1}_{1\times 8} & \mathbf{0}_{1\times 3} \\ & f_3(z) & \end{bmatrix} u, \tag{3.14}$$

where

- y_1 is the sum of the simple moving averages over the past 60 min for zones 1 to 5, i.e., $f_1(z) = \frac{1}{20}\sum_{k=0}^{19} z^{-k}$,
- y_2 is $\sum_{i=5}^{12} f_2 u_i$, with f_2 a low-pass filter with Gaussian shaped finite impulse response of length 20, obtained in MATLAB as `gaussdesign(0.5,2,10)`.
- y_3 is the scalar output of a 1×15 MISO filter f_3 designed to forecast at each period t the average total number of events per time-period that will occur in the whole building during the window $[t + 60 \text{ min}, t + 90 \text{ min}]$. This filter was constructed by identifying an ARMAX model (Ljung 1998) between the 15 inputs (plus a scalar white noise) and the desired output, with the calibration done using one part of the dataset. The model chosen takes the form

$$y_{3,t} = \sum_{i=1}^{4} a_i y_{3,t-i} + b_0 u_t + b_1 u_{t-1} + e_t + c_1 e_{t-1},$$

where $a_1, \ldots, a_4$ and b_0, b_1 are scalar and row vectors respectively forming the filter f_3, c_1 is a scalar and e_t is a zero-mean standard white noise input postulated by the ARMAX model for system identification purposes.

Figure 3.4 shows sample paths over a 72 h period of the 2nd and 3rd outputs of the desired filter and of its $(\ln 5, 0.05)$-differentially private approximation obtained using the ZFE MIMO mechanism with diagonal pre-filter. The 15 optimal pre-filters were obtained approximately via least-squares fit of $\sqrt{|F_i(e^{j\omega})|_2}$ with negligible approximation error (using MATLAB's function yulewalk implementing the Yule–Walker method (Stoica and Moses 2005), rather than true spectral factorization as mentioned in Theorem 3.2. Among other things, one can notice that the noise remaining on each output can have quite different characteristics depending on the desired filter F, with the post-filter FG^{-1} removing more high-frequency components on the second output than on the third. Referring to (3.11) for the performance achieved with these diagonal pre-filters and the lower bound (3.13), we obtain for the filter F designed

$$\frac{1}{2\pi} \int_{-\pi}^{\pi} \sum_{i=1}^{15} \rho_i |F_i(e^{j\omega})|_2 \, d\omega \approx 2.2 \times \frac{1}{2\pi} \int_{-\pi}^{\pi} \|F(e^{j\omega})R\|_* d\omega.$$

Hence, restricting the ZFE mechanism to diagonal pre-filters degrades the root-mean-square error (RMSE) by a multiplicative factor of at most 2.2 in this case.

References

Cortés J et al (2016) Differential privacy in control and network systems. In: Proceedings of the 55th conference on decision and control, Las Vegas, NV

Horn RA, Johnson CR (2012) Matrix analysis, 2nd edn. Cambridge University Press, Cambridge

Ivanov YA et al (2007) Visualizing the history of living spaces. IEEE Trans Vis Comput Graphics 13(6):1153–1160

Kailath T, Sayed AH, Hassibi B (2000) Linear estimation. Prentice Hall, Upper Saddle River

Le Ny J, Mohammady M (2018) Differentially private MIMO filtering for event streams. IEEE Trans Autom Control 63(1):145–157

Li C, Miklau G (2012) An adaptive mechanism for accurate query answering under differential privacy. In: Proceedings of the conference on very large databases (VLDB), Istanbul, Turkey

Ljung L (1998) System identification: theory for the user. Information and system sciences. Prentice Hall, Uppder Saddle River

Salz J (1985) Digital transmission over cross-coupled linear channels. AT&T Tech J 64(6):1147–1159

Sankar L, Rajagopalan SR, Poor HV (2010) A theory of privacy and utility in databases. In: Proceedings of the IEEE international symposium on information theory

Stoica P, Moses RL (2005) Spectral analysis of signals. Prentice Hall, Upper Saddle River

Tanaka T et al (2017) Semidefinite programming approach to gaussian sequential rate-distortion trade-offs. IEEE Trans Autom Control 62(4)

Wren C et al (2007) The MERL motion detector dataset: 2007 workshop on massive datasets. Technical report TR2007-069. Mitsubishi Electric Research Laboratories

Yang J, Roy S (1994) On joint transmitter and receiver optimization for multiple-input-multiple-output (MIMO) transmission systems. IEEE J Commun 42(12):3221–3231

Chapter 4
Differentially Private Filtering for Stationary Stochastic Collective Signals

4.1 Introduction

[1]Chapter 3 described the zero-forcing equalization (ZFE) mechanism, a two-stage mechanism for differentially private filtering of collective signals, as defined in Sect. 2.3.1. In the ZFE mechanism, the privacy-preserving noise is added at the output of the first stage, a signal shaping block, while the second stage tries to cancel the effect of the first stage by simply inverting it, in order to estimate the output signal of interest. This design has the advantage of not requiring any model of the privacy-sensitive input signal for its implementation. On the other hand, in many situations of interest, we might have some public knowledge about this input signal, e.g., in the form of a statistical or physical model. It is then natural to ask how we can leverage such knowledge to design differentially private mechanisms with better performance than the ZFE mechanism. In practice, this is done by choosing a more appropriate structure than an inverse filter for the second stage of the mechanism.

In this chapter, we start addressing the question of how to leverage signal models in the design of differentially private filters, by considering the same set-up as for the ZFE mechanism, but assuming now that the input collective signals are wide-sense stationary with known statistics. In this case, one can replace the inverse filter of the second stage by a Wiener filter for example, which takes the statistical model into account to produce a more accurate output. Indeed, focusing here on SISO systems to simplify the discussion, one issue with zero-forcing equalizers is the noise amplification at frequencies where the input block frequency response $|G(e^{j\omega})|$ is small, due to the presence of the inverse filter G^{-1} in the second stage (Proakis 2000). This issue is mitigated in the optimal ZFE mechanism, since in this case the amplification is compensated by the fact that $|F(e^{j\omega})|$ and $|G(e^{j\omega})|$ given in Theorem 3.2 are both small at the same frequencies. Nonetheless, in general using

---

[1]Some of the text in Sect. 4.3 of this chapter is reprinted, with permission, from Le Ny and Mohammady (2018) (© [2018] IEEE).

© The Author(s), under exclusive license to Springer Nature Switzerland AG 2020
J. Le Ny, *Differential Privacy for Dynamic Data*,
SpringerBriefs in Control, Automation and Robotics,
https://doi.org/10.1007/978-3-030-41039-1_4

a Wiener filter as the post-filter H of Fig. 3.2 improves on the ZFE mechanism by adapting the filter's response to the noise power and input signal power at each frequency.

4.2   Linear Minimum Mean Square (Wiener) Smoothing and Filtering

Appendix A presents some basic definitions and notations from statistical signal processing necessary for our discussion. Here we recall some elements of Wiener's theory for the linear minimum mean square estimation of discrete-time stationary stochastic processes. More details can be found in Kailath et al. (2000), Chap. 7 for example. Consider two jointly WSS discrete-time signals y and v with zero mean, of which only v is observed and should be used to produce an estimate of y. More specifically, we wish to produce an estimate $\hat{y}$ of y minimizing the MSE, but with the restriction that $\hat{y}$ be linear in v. In other words, $\hat{y}$ should be the output of an linear system H when the input signal is v. At first, we do not restrict H to be causal or time-invariant, so that we can write, for all t,

$$\hat{y}_t = \sum_{\tau=-\infty}^{\infty} H_{t,\tau} v_\tau,$$

where $H_{t,\tau}$ is the system's response at time t to an impulse applied at time τ. We wish to choose $H_{t,\tau}$ to achieve at each period t the minimum mean-squared error (MMSE), i.e., minimize $\mathbb{E}\left[|y_t - \hat{y}_t|_2^2\right]$.

Using standard linear least squares minimization arguments (Kailath et al. 2000), one can show that due to the stationarity of v, y, the filter H must be time-invariant and that the following relationship between correlation matrices must hold

$$R_t^{yv} = \sum_{\tau=-\infty}^{\infty} H_{t-\tau} R_\tau^v = \sum_{\tau=-\infty}^{\infty} H_\tau R_{t-\tau}^v, \quad \forall -\infty < t < \infty, \tag{4.1}$$

where H_t denotes the impulse response of H. Hence, to obtain the linear minimum mean-squared error (LMMSE) estimator, the second order (joint) statistics of the processes y, v, are assumed known. Taking the z-transform of (4.1), we obtain a formula for the LMMSE estimator's transfer function as

$$H(z) = P_{yv}(z) P_v(z)^{-1}. \tag{4.2}$$

This transfer function is rational if the z-spectra are, an assumption that we make in the following for all signals to simplify the discussion. We also assume for simplicity that $P_v(e^{j\omega})$ is positive definite, for all $\omega \in (-\pi, \pi]$. The system H in (4.2) is not necessarily causal however and in general the estimate $\hat{y}_t$ at time t can depend on the values for v at all times, including arbitrarily far in the future. Hence, we refer to the system with transfer function (4.2) as the Wiener smoother. The MMSE (over linear systems) achieved by the Wiener smoother can be obtained explicitly as follows.

First, note that $\hat{y}$ is WSS because v is and $\hat{y} = Hv$. We have

$$P_{\hat{y}}(e^{j\omega}) = H(e^{j\omega})P_v(e^{j\omega})H(e^{j\omega})^* = P_{yv}(e^{j\omega})P_v(e^{j\omega})^{-1}P_{yv}(e^{j\omega})^*,$$

where (4.2) was used to obtain the final expression. Again, standard linear least squares minimization argument showing that the residual $y - \hat{y}$ must be orthogonal to $\hat{y}$ leads to the following expression for the error (a Pythagorean theorem)

$$\mathbb{E}[|y_t - \hat{y}_t|^2] = \mathbb{E}[|y_t|^2] - \mathbb{E}[|\hat{y}_t|^2] = R_0^y - R_0^{\hat{y}} = \frac{1}{2\pi}\int_{-\pi}^{\pi} P_y(e^{j\omega}) - P_{\hat{y}}(e^{j\omega})d\omega$$

$$= \frac{1}{2\pi}\int_{-\pi}^{\pi} P_y(e^{j\omega}) - P_{yv}(e^{j\omega})P_v(e^{j\omega})^{-1}P_{yv}(e^{j\omega})^* d\omega. \qquad (4.3)$$

For online implementations, i.e., when the input data v arrives continuously, the Wiener smoother can sometimes be implemented approximately by introducing a delay sufficient for the impulse response of H to become negligible. In other words, we produce at any time t an estimate $\hat{y}_{t-\tau}$, for some fixed delay τ, based on the values v_s for all $s \leq t$. An appropriate value of τ depends on the specific joint statistics of y, v for the problem considered, and on the tolerable degree of approximation. However, a more rigorous approach is to directly design a *causal* LMMSE estimator, i.e., a Wiener filter rather than a Wiener smoother, which estimates $\hat{y}_t$ at each time using only the values v_s for $s \leq t$. The Wiener filter is also time-invariant, and its impulse response is defined as the solution of the Wiener–Hopf equations

$$R_t^{yv} = \sum_{\tau=-\infty}^{\infty} H_{t-\tau}R_\tau^v = \sum_{\tau=-\infty}^{\infty} H_\tau R_{t-\tau}^v, \quad \forall 0 \leq t < \infty, \qquad (4.4)$$

$$H_t = 0, \quad \forall t < 0, \qquad (4.5)$$

where, compared to (4.1), the relation (4.4) between correlation matrices only holds for $t \geq 0$, and moreover we have the causality constraint (4.5).

The solution of the causal Wiener filtering problem is more involved than for the Wiener smoother, especially in the multidimensional case, since a spectral factorization step is involved. First, we define the following notation. If $F(z)$ is an analytic function in some annulus that contains the unit circle, with expansion $F(z) = \sum_{i=-\infty}^{\infty} F_t z^{-t}$, then the causal part of the function is denoted

$$\{F(z)\}_+ := \sum_{t=0}^{\infty} F_t z^{-t},$$

and the anticausal part is $\{F(z)\}_- := F(z) - \{F(z)\}_+$. When $F(z)$ is rational and proper, one can compute $\{F(z)\}_+$ in practice using a partial fraction expansion, see Kailath et al. (2000), Sect. 7.6. Under our assumptions for P_v, there exists a positive definite matrix R_e and a rational canonical spectral factor $L(z)$ such that

$$P_v(z) = L(z)\ R_e\ L(z^{-1})^T,$$

where $L(z)$ and $L^{-1}(z)$ are both analytic on and outside the unit circle (i.e., are stable and minimum-phase), and $L(\infty) = I$, the identity matrix. The solution of the Wiener–Hopf equation can then be written, with the notation $A^{-T} := (A^T)^{-1}$,

$$H(z) = \left\{ P_{yv}(z)L(z)^{-T} \right\}_+ R_e^{-1} L^{-1}(z). \tag{4.6}$$

4.3 Two-Stage Differentially Private LMMSE Estimators

In this section, we return to the differentially private filtering problem considered in Chap. 3, but we now assume some publicly known information about the privacy-sensitive input signal u on Fig. 3.2, in the form of a statistical model. Namely, we assume that u is WSS with known second-order statistics, i.e., this signal has a known constant mean vector μ and matrix-valued autocorrelation sequence R_t^u, following the notation introduced in Appendix A. We continue to assume that the prescribed filter F on Fig. 3.2 is an LTI (stable) system, with zero initial conditions. Hence, the desired output signal y is also WSS. In the following, we assume without loss of generality that μ is zero. Indeed, if this is not the case we can write $u = \mu + u_1$ and $y = F(1)\mu + Fu_1$, with u_1 now a centered signal with autocorrelation matrix $R_t^{u_1} = R_t^u - \mu\mu^T$, and the problem reduces to producing a differentially private signal approximating Fu_1. The z-spectrum matrix of u is denoted $P_u(z)$ and is assumed to have rational entries and to be positive definite on the unit circle, i.e., $P_u(e^{j\omega}) \succ 0$, for all $\omega \in [-\pi, \pi)$.

Recall that for the differentially private filtering architecture of Fig. 3.2, the role of the system H is to estimate the signal y from the perturbed signal v coming from the pre-filter G. For w is a WSS signal uncorrelated with u with z-spectrum P_w, the signal v on Fig. 3.2 is WSS with a z-spectrum equal to

$$P_v(z) = G(z)P_u(z)G(z^{-1})^T + P_w(z), \tag{4.7}$$

see Appendix A. Similarly, the signal y is WSS and the cross z-spectrum P_{yv} is

$$P_{yv}(z) = F(z)P_u(z)G(z^{-1})^T. \tag{4.8}$$

From Sect. 4.2, and in particular (4.2) and (4.6), knowledge of these z-spectra allows us to implement for H a linear smoother or filter minimizing among linear systems the MSE between y and $\hat{y}$ on Fig. 3.2. Although other performance criteria than MSE could be appropriate to design H, such as maximum likelihood estimation, designing linear estimators minimizing the MSE typically leads to relatively tractable problems and is in some sense compatible with the performance measure (3.1). Hence, in this chapter, we study mechanisms where H is designed as a LMMSE estimator, which we call LMMSE mechanisms. This fixes the structure of H, and it then remains to optimize over the choice of G, following the general methodology outlined in Sect. 3.2.

There are still two main obstacles to the design of optimal LMMSE mechanisms. First, for systems F with multiple inputs, as for the ZFE mechanism of Chap. 3, we need to add white noise w on Fig. 3.2 proportional to the sensitivity of the filter G, for which we do not have a tractable formula. Hence, as in Sect. 3.4.2, we resort to optimizing over diagonal prefilters G only. Second, even for SISO systems, the expression of the MSE (3.1) is a complicated function of G when H is a causal Wiener filter, given by (4.6). Hence, we focus first on scenarios where H is a Wiener smoother, i.e., not necessarily causal, with the corresponding simpler expression (4.2). For the resulting simplified architecture, we can express the estimation performance analytically as a function of G. Optimizing this performance measure provides the best diagonal prefilter for a Wiener *smoother* H, and a lower bound on the performance achievable with any diagonal prefilter followed by a Wiener *filter* H.

For specific implementations of *causal* LMMSE mechanisms, two natural choices are then discussed. First, we can take the pre-filter G obtained from the computation of the lower bound assuming a Wiener smoother, and implement instead for H a *causal* Wiener filter, or perhaps a slightly non-causal filter if it can be implemented by introducing a delay that is tolerable for a specific application. This mechanism will not attain the lower bound in general, since the bound and corresponding G were obtained by removing the causality constraint on H. Another possibility is to take G as in the ZFE mechanism, and simply replace $H = FG^{-1}$ by a Wiener filter. This choice has the advantage of always improving on the ZFE mechanism, and tends to perform well in practice. For any specific implementation with diagonal prefilters, the lower bound on performance can be computed to provide an indication of how far the design might be from the optimal one.

4.3.1 Lower Bound on MSE for Diagonal Pre-filters

From (4.2), the (non-causal) Wiener smoother H has the transfer function $H(z) = P_{yv}(z)P_v(z)^{-1}$. According to Theorem 2.7, for G diagonal we can take the privacy-preserving noise w to be white and Gaussian (independent of u) with covariance $\sigma^2 I_m$, where $\sigma^2 = \kappa_{\delta,\varepsilon}^2 \|GR\|_2^2$ and $R = \text{diag}(\rho_1, \ldots, \rho_m)$ for the adjacency relation (2.5). That is, P_w in (4.7) is constant and equal to $P_w(z) = \sigma^2 I_m$. Hence,

$$H(z) = F(z)P_u(z)G(z^{-1})^T \left(G(z)P_u(z)G(z^{-1})^T + \kappa_{\delta,\varepsilon}^2 \|G\|_2^2 I_m \right)^{-1}. \tag{4.9}$$

The MSE (3.1) corresponding to such an architecture (diagonal prefilter followed by a Wiener smoother) is expressed by (4.3). In our case,

$$e_{mse}^{LMMSE}(G) = \frac{1}{2\pi} \int_{-\pi}^{\pi} \text{Tr}(P_y(e^{j\omega}) - P_{\hat{y}}(e^{j\omega}))d\omega$$

with

$$P_{\hat{y}}(e^{j\omega}) = H(e^{j\omega})P_v(e^{j\omega})H(e^{j\omega})^* = P_{yv}(e^{j\omega})P_v(e^{j\omega})^{-1}P_{yv}(e^{j\omega})^*$$
$$= FP_uG^*(\sigma^2 I_m + GP_uG^*)^{-1}GP_uF^*,$$

where on the last line and in the following we often omit the argument $e^{j\omega}$ of the transfer functions, to simplify the notation. Let us denote $P_u(e^{j\omega}) = \Delta(e^{j\omega})\Delta(e^{j\omega})$, where $\Delta(e^{j\omega}) \succeq 0$ is the principal square root of $P_u(e^{j\omega})$. We have then

$$P_y - P_{\hat{y}} = FP_uF^* - P_{\hat{y}}$$
$$= F\Delta(I_m - \Delta G^*(\sigma^2 I_m + G\Delta\Delta G^*)^{-1}G\Delta)\Delta F^*$$
$$= F\Delta\left(I_m + \frac{1}{\sigma^2}\Delta G^*G\Delta\right)^{-1}\Delta F^*,$$

where the last expression is obtained using the matrix inversion lemma. Finally, defining $\tilde{G}(e^{j\omega}) := \frac{1}{\|GR\|_2}G(e^{j\omega})R$, we obtain the expression

$$e_{mse}^{LMMSE}(\tilde{G}) = \frac{1}{2\pi}\int_{-\pi}^{\pi} \text{Tr}\left[F\Delta Z^{-1}\Delta F^*\right]d\omega \qquad (4.10)$$

$$\text{with } Z = \left(I_m + \frac{1}{\kappa_{\delta,\varepsilon}^2}\Delta R^{-1}\tilde{G}^*\tilde{G}R^{-1}\Delta\right),$$

where again the arguments $e^{j\omega}$ were omitted. The objective (4.10) should be minimized over all transfer functions $\tilde{G}$ satisfying the constraint

$$\|\tilde{G}\|_2^2 = \frac{1}{2\pi}\int_{-\pi}^{\pi} \text{Tr}(\tilde{G}(e^{j\omega})^*\tilde{G}(e^{j\omega}))d\omega = 1. \qquad (4.11)$$

To obtain a lower bound on performance (for diagonal pre-filters), we now minimize the performance measure (4.10) over the choice of diagonal pre-filters G satisfying (4.11). First, in the case where $P_u(e^{j\omega})$ is positive definite and diagonal for all ω, i.e., the different input signals are uncorrelated, we have in fact an allocation problem whose solution is of the "waterfilling type" (Palomar and Fonollosa 2005). Namely, denote

$$P_u(e^{j\omega}) = \text{diag}(p_1(e^{j\omega}), \ldots, p_m(e^{j\omega}))$$

and

$$X(e^{j\omega}) = \tilde{G}(e^{j\omega})^*\tilde{G}(e^{j\omega}) = \text{diag}(x_1(e^{j\omega}), \ldots, x_m(e^{j\omega})),$$

with $x_i(e^{j\omega}) = |\tilde{G}_{ii}(e^{j\omega})|^2$. Omitting the expression $e^{j\omega}$ in the integrals for clarity, (4.10) and (4.11) read

$$\min_{x(\cdot)} \quad \frac{1}{2\pi} \int_{-\pi}^{\pi} \sum_{i=1}^{m} \frac{1}{\frac{1}{p_i} + \frac{x_i}{\kappa_{\delta,\varepsilon}^2 \rho_i^2}} |F_i|_2^2 \, d\omega$$

$$\text{s.t.} \quad \frac{1}{2\pi} \int_{-\pi}^{\pi} \sum_{i=1}^{m} x_i \, d\omega = 1, \quad x_i(e^{j\omega}) \geq 0, \forall \omega, i,$$

and the solution to this convex problem is

$$x_i(e^{j\omega}) = \max \left\{ 0, \frac{\kappa_{\delta,\varepsilon} \rho_i |F_i(e^{j\omega})|_2}{\lambda} - \frac{\kappa_{\delta,\varepsilon}^2 \rho_i^2}{p_i(e^{j\omega})} \right\}, \tag{4.12}$$

where $\lambda > 0$ is adjusted so that the solution satisfies the equality constraint (4.11) (note that since (4.12) is not smooth we would have to approximate this solution if we wanted to implement the corresponding G with a finite-dimensional filter). Problems of this type are discussed in the communication literature on joint transmitter-receiver optimization (Salz 1985; Yang and Roy 1994), which is not too surprising in view of our approximation setup on Fig. 3.2.

When $P_u(e^{j\omega})$ is not diagonal, one can obtain a solution arbitrarily close to the minimum one using semidefinite programming. First, we discretize the optimization problem at the set of frequencies $\omega_q = \frac{q\pi}{N}, q = 0, \ldots, N$. Note that all functions are even functions of ω, hence we can restrict out attention to the interval $[0, \pi]$. Then, we define the $m(N + 1)$ variables $x_{iq} = x_i(e^{j\omega_q})$, with $x_{iq} \geq 0$, and $X_q = \text{diag}(x_{1q}, \ldots, x_{mq})$. Using the trapezoidal rule to approximate the integrals, we obtain the following optimization problem

$$\min_{\{X_q, M_q\}_{0 \leq q \leq N}} \quad \frac{1}{2N} \sum_{q=0}^{N-1} \text{Tr} \left[M_q + M_{q+1} \right] \tag{4.13}$$

$$\text{s.t.} \quad \begin{bmatrix} M_q & F_q \Delta_q \\ \Delta_q F_q^* & I_m + \tilde{\Delta}_q X_q \tilde{\Delta}_q^* \end{bmatrix} \succeq 0, \quad 0 \leq q \leq N, \tag{4.14}$$

$$\frac{1}{2N} \sum_{q=0}^{N-1} \text{Tr}[X_q + X_{q+1}] = 1, \quad \text{and } X_q \succeq 0, \quad 0 \leq q \leq N,$$

where $F_q := F(e^{j\omega_q})$, $\Delta_q := \Delta(e^{j\omega_q})$ and $\tilde{\Delta}_q = \frac{1}{\kappa_{\delta,\varepsilon}} \Delta_q R^{-1}$. Note that (4.14) is equivalent to $M_q \succeq F_q \Delta_q \left(I_m + \tilde{\Delta}_q X_q \tilde{\Delta}_q^* \right)^{-1} \Delta_q F_q^*$ by taking the Schur complement. The optimization problem (4.13), (4.14) is a semidefinite program (SDP) (Boyd et al. 1994). It involves a cost function that is linear in the (matrix) variables subject to constraints that are either linear or specify that certain matrices must be positive semidefinite, i.e., linear matrix inequalities (LMIs). SDPs of moderate size can be solved relatively efficiently. This allows us to discretize the interval $[0, \pi]$ sufficiently finely and obtain by solving the corresponding SDP a value providing a lower bound on the performance achievable by any LMMSE mechanism using a diagonal pre-filter G and a Wiener filter or smoother H.

4.3.2   Pre- and Post-filter Design

We discuss here two possible differentially private mechanisms to try to approach the performance lower bound computed in Sect. 4.3.1. First, one can simply take the square root pre-filter G of Theorem 3.2 for the ZFE mechanism, or a rational approximation of it, followed by a (causal) Wiener filter H. With $G(z)$ rational and with our standing assumption that P_u is rational, then P_v is rational as well, and note from (4.7) that $P_v(e^{j\omega}) \succ 0$ for all $-\pi \leq \omega < \pi$. Hence, $P_v(z)$ has a canonical spectral factorization $P_v(z) = L(z) P_e L(z^{-1})^T$, and the causal Wiener filter is given by (4.6). This architecture improves on the ZFE mechanism if the statistical assumptions on u are satisfied. Another possible pre-filter G is the one obtained from solving the SDP above. The transfer functions $\tilde{G}_{ii}$ (and hence G_{ii}) of the filter $\tilde{G}$ can then be obtained by interpolation of their squared magnitude $x_i(e^{j\omega})$ from the variables x_{iq} and m spectral factorizations. The resulting filter G would only be optimal (among diagonal filters) if we were using a Wiener smoother H, i.e., if we relaxed the causality constraint on H. By implementing a causal Wiener filter H, it is in fact possible that this architecture performs worse than the previous one for specific problems.

Figure 4.1 shows an example of sample paths obtained using the ZFE mechanism and the LMMSE mechanism for a filtering problem detailed in Le Ny and

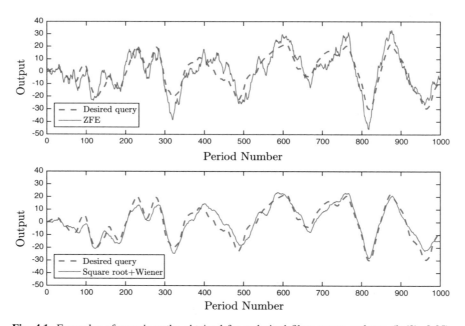

Fig. 4.1 Examples of sample paths obtained for a desired filter output and two $(\ln(3), 0.05)$-differentially private approximations, using the ZFE mechanism (top, empirical RMSE $= 7.1$), and the same mechanism with the inverse post-filter replaced by a (causal) Wiener filter (bottom, empirical RMSE $= 4.5$). © [2018] IEEE. Reprinted, with permission, from Le Ny and Mohammady (2018)

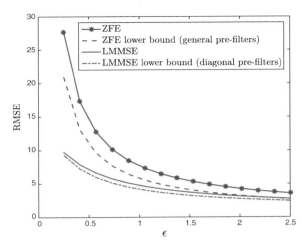

Fig. 4.2 Performance comparison (RMSE) for $(\varepsilon, 0.05)$-differentially private ZFE and LMMSE mechanisms (both using the same square root pre-filter), as a function of ε, for the problem described in Le Ny and Mohammady (2018), Sect. VI.B. Significant performance improvement can be obtained by leveraging a statistical model of the privacy-sensitive input signal, especially in the high privacy regime (small values of ε). The lower bounds are useful in assessing how far a particular design is from the optimum one could potentially achieve in a larger class of architectures. © [2018] IEEE. Reprinted, with permission, from Le Ny and Mohammady (2018)

Mohammady (2018), Sect. VI.B. In this example, we assume that the input signal u is described by the evolution of a Markov chain in steady-state, from which one can compute the PSD P_u. The inverse filter in H can then be replaced by a causal Wiener filter, which gives the apparent better accuracy for the output shown on the figure.

For these two mechanisms providing (ε, δ)-differential privacy and the same problem, Fig. 4.2 shows the RMSE obtained for $\delta = 0.05$ and different values of ε. We also plot the performance lower bounds for ZFE mechanisms with *general* pre-filters (Theorem 3.3) and for LMMSE mechanisms for any *diagonal* pre-filter (solution of (4.13)). We can see that the performance of ZFE mechanisms degrades quickly ε becomes small, i.e., when a significant amount of privacy preserving noise is introduced, even if we could optimize over general pre-filters rather than just diagonal ones. Recall in particular from (3.13), (3.11) that the dependance on the privacy parameters ε, δ of both the diagonal ZFE mechanism performance and of the lower bound is captured completely by the multiplicative factor $\kappa_{\delta,\varepsilon}$. This is not the case for the LMMSE mechanism however, since the Wiener filter takes the variance of the privacy noise into account. Moreover, for the LMMSE mechanism, we see in this specific example that little performance improvement can be obtained by further optimizing the diagonal pre-filter or even by replacing the post-filter H by a non-causal Wiener smoother.

References

Boyd S et al (1994) Linear matrix inequalities in system and control theory. Studies in applied mathematics. SIAM, Philadelphia

Kailath T, Sayed AH, Hassibi B (2000) Linear estimation. Prentice Hall, Upper Saddle River

Le Ny J, Mohammady M (2018) Differentially private MIMO filtering for event streams. IEEE Trans Autom Control 63(1):145–157

Palomar DP, Fonollosa JR (2005) Practical algorithms for a family of waterfilling solutions. IEEE Trans Signal Process 53(2):686–695

Proakis J (2000) Digital communications, 4th edn. McGraw-Hill, New York

Salz J (1985) Digital transmission over cross-coupled linear channels. AT&T Tech J 64(6):1147–1159

Yang J, Roy S (1994) On joint transmitter and receiver optimization for multiple-input-multiple-output (MIMO) transmission systems. IEEE J Commun 42(12):3221–3231

Chapter 5
Differentially Private Kalman Filtering

5.1 Introduction

[1]In this chapter, we continue discussing the design of model-based differentially private filters. Whereas in Chap. 4 we assumed that the privacy-sensitive signals could be modeled as stationary signals with publicly known second-order statistics, we consider here scenarios where these signals can be modeled as the output of a linear finite-dimensional system with publicly known parameters. This relaxes in particular the stationarity assumption. The mathematical model could capture for example known physical laws that govern the behavior of the input signals, such as a kinematic model relating successive position measurements obtained from an individual to his or her velocity. The goal is to use the collected privacy-sensitive signals to estimate the state of the modeled system, or perhaps some aggregate linear function of this state. For example, we might want to publish a real-time estimate of the average speed of a group of drivers who send their individual locations over time, by leveraging a kinematic model for each car. As in the previous chapter, the availability of a model should improve the estimation performance, but we still need to provide a differential privacy guarantee for the collected signals. In the absence of privacy constraint, the best linear state estimator minimizing the MSE performance measure is the Kalman filter, which is also optimal among all filters under an additional assumption that the process and measurement noises of the model have Gaussian distributions. In this chapter, we adapt the Kalman filter in various ways in order to produce differentially private real-time statistics in scenarios involving the processing of either individual or collective signals, as defined in Sect. 2.3.1.

---

[1]Some of the text in the Sects. 5.4.1, 5.4.2 and 5.4.4 of this chapter is reprinted, with permission, from Le Ny and Pappas (2014) (© [2014] IEEE).

© The Author(s), under exclusive license to Springer Nature Switzerland AG 2020
J. Le Ny, *Differential Privacy for Dynamic Data*,
SpringerBriefs in Control, Automation and Robotics,
https://doi.org/10.1007/978-3-030-41039-1_5

5.2   Differentially Private Kalman Filtering Problems

In this section, we state differentially private Kalman filtering problems both for individual and collective data streams.

5.2.1   Individual Data Streams

For the first type of scenarios, we consider a set of n private individual data streams $\{u_{i,t}\}_{0 \leq t \leq T}$, for $1 \leq i \leq n$, with $u_{i,t} \in \mathbb{R}^{d_i}$, which could originate from n individuals for example. Let $d := \sum_{i=1}^{n} d_i$. Both the cases $T < \infty$ and $T = \infty$ are of interest. We assume a publicly known mathematical model for these signals, namely, we posit that they can be explained as the outputs of n independent linear finite-dimensional dynamical systems with state-space representations

$$x_{i,t+1} = A_{i,t} x_{i,t} + B_{i,t} w_{i,t}, \ t = 0, 1, ..., T - 1, \tag{5.1}$$

$$u_{i,t} = C_{i,t} x_{i,t} + D_{i,t} w_{i,t}, \quad t = 0, 1, ..., T, \tag{5.2}$$

for $1 \leq i \leq n$, where $x_{i,t} \in \mathbb{R}^{m_i}$, $w_{i,t}$ is a zero-mean standard white noise process (sequence of iid random variables with covariance $\mathbb{E}[w_{i,t} w_{i,t'}] = \delta_{t-t'} I$), and the initial condition $x_{i,0}$ is a random variable with mean $\bar{x}_{i,0}$ and covariance $\bar{\Sigma}_{i,0}$, independent of the noise process w_i. We assume for simplicity that the matrices $D_{i,t}$ are full row rank. When $T = \infty$, we also assume that the matrices do not depend on time and denote them A_i, B_i, C_i, D_i. In that case, for all $1 \leq i \leq n$, we assume that the pairs (A_i, C_i) are detectable and the pairs (A_i, B_i) are stabilizable.

The signals u_i are sent to a trusted data aggregator, whose goal is to release a new signal that asymptotically minimizes the MSE with respect to a linear combination of the individual states. That is, the quantity of interest to be estimated at each period is $z_t = \sum_{i=1}^{n} L_{i,t} x_{i,t}$, where $L_{i,t}$ are given matrices, and we are looking for a causal estimator $\hat{z}$ constructed from the signals u_i, $1 \leq i \leq n$, minimizing the quantity

$$e_{mse}^{KF,T} := \frac{1}{T} \sum_{t=0}^{T-1} \mathbb{E}\left[\| z_t - \hat{z}_t \|_2^2 \right], \tag{5.3}$$

when $T < \infty$, or the limit

$$e_{mse}^{KF,\infty} := \lim_{T \to \infty} \sup e_{mse}^{KF,T} \tag{5.4}$$

in the infinite stream case (in this case, the matrices $L_{i,t}$ are also assumed time-invariant and denoted L_i). The data $\bar{x}_{i,0}$, $\bar{\Sigma}_{i,0}$, $A_{i,t}$, $B_{i,t}$, $C_{i,t}$, $D_{i,t}$, $L_{i,t}$, $1 \leq i \leq n$, are considered to be public information. In the absence of privacy constraint, the linear minimum mean squared error (LMMSE) estimator is $\hat{z}_t = \sum_{i=1}^{n} L_{i,t} \hat{x}_{i,t}$, with

$\hat{x}_{i,t}$ provided by the Kalman filter, denoted $\mathcal{K}_i$ in the following, estimating the state of system i from u_i (Anderson and Moore 2005), i.e., $\hat{z} = \sum_{i=1}^{n} L_{i,t}(\mathcal{K}_i u_i)_t$. This filter is time-varying in general, but in the infinite stream scenario the time-invariant Kalman filter also optimizes the steady-state performance $e_{mse}^{KF,\infty}$. If moreover the noise w is assumed to be a Gaussian process, then the Kalman filter is a MMSE estimator, i.e., it minimizes the MSE among all filters, not just the linear ones (Anderson and Moore 2005).

We now impose a differential privacy constraint for the estimation task just described, in order to protect the input signals $u = [u_1^T, \ldots, u_n^T]^T$. Since we are concerned here with individual data streams, an appropriate adjacency relation on the signals u is given by (2.3) for example. Such an adjacency relation has the advantage of being defined solely in terms of the physical (measured) input data. Alternatively, we could aim to protect the state signals of the dynamic model. Let $x = [x_1^T, \ldots, x_n^T]^T$ denote the global state signal for the model (5.1). Assume for instance that the mechanism is required to guarantee differential privacy for a subset $\mathscr{S}_i \subset \{1, \ldots, m_i\}$ of the coordinates of the state trajectory x_i. Let the selection matrix S_i be the diagonal matrix with $[S_i]_{jj} = 1$ if $j \in \mathscr{S}_i$, and $[S_i]_{jj} = 0$ otherwise. Hence $S_i v$ sets the coordinates of a vector v that do not belong to the set $\mathscr{S}_i$ to zero. Fix a constant $\rho \in \mathbb{R}_+$. The adjacency relation could then be defined as

$$\text{Adj}_{\mathscr{S}}^{\rho}(x, x') \text{ iff for some } i, \ \|S_i x_i - S_i x_i'\|_2 \leq \rho, \tag{5.5}$$
$$(I - S_i)x_i = (I - S_i)x_i', \text{ and } x_j = x_j' \text{ for all } j \neq i.$$

In words, two adjacent global state trajectories differ by the values of a single participant, say i. Moreover, the range in energy variation in the signal $S_i x_i$ of participant i is constrained to be at most ρ^2. A more general version of this set-up consists in taking each S_i to be a projection matrix, i.e., a square matrix such that $S_i^2 = I$. This is the only property beyond (5.5) that will be required in the following discussion.

If the state vector x_i has a clear physical interpretation for user i, using an adjacency relation such as (5.5) might be meaningful. The main drawback compared to placing the adjacency relation on u is that (5.1) is only a mathematical model, and hence the privacy guarantee becomes dependent on the accuracy of this model. In particular, if (5.1) is only a rough approximation of the behavior of a real state signal that we would like to protect and of the measurement process, then it can be unclear what concrete guarantees the adjacency relation (5.5) offers. On the other hand, an advantage is that the "natural" measurement noise appearing in (5.2) can contribute to preserving privacy, assuming again that we trust the mathematical measurement model, and in particular our knowledge of the measurement noise distribution.

With the adjacency relation now defined, we aim to design a mechanism M producing a signal Mu approaching as closely as possible the LMMSE estimate $\hat{z}$, while maintaining differential privacy for the chosen adjacency relation.

Example 5.1 Consider the following simplified traffic monitoring system. There are n participating vehicles traveling on a straight road segment. Vehicle i, for $1 \leq i \leq n$, is represented by its state $x_{i,t} = [\xi_{i,t}, \dot{\xi}_{i,t}]^T$, with ξ_i and $\dot{\xi}_i$ its position and velocity respectively. This state evolves as a second-order system with unknown random acceleration inputs

$$x_{i,t+1} = \begin{bmatrix} 1 & T_s \\ 0 & 1 \end{bmatrix} x_{i,t} + \sigma_1 \begin{bmatrix} T_s^2/2 & 0 \\ T_s & 0 \end{bmatrix} w_{i,t},$$

where T_s is the sampling period, $w_{i,t}$ is a standard white Gaussian noise, and $\sigma_1 > 0$. Assume for simplicity that the noise signals w_j for different vehicles are independent. The traffic monitoring service collects GPS measurements from the vehicles, i.e., receives noisy readings of the positions at the sampling times

$$u_{i,t} = \begin{bmatrix} 1 & 0 \end{bmatrix} x_{i,t} + \sigma_2 \begin{bmatrix} 0 & 1 \end{bmatrix} w_{i,t}, \text{ with } \sigma_2 > 0.$$

The purpose of the traffic monitoring service is to continuously provide an estimate of the traffic flow velocity on the road segment, which is approximated by releasing at each sampling period an estimate of the average velocity of the participating vehicles, i.e., of the quantity

$$z_t = \frac{1}{n} \sum_{i=1}^{n} \dot{\xi}_{i,t}. \tag{5.6}$$

With a larger number of participating vehicles, the sample average (5.6) represents the traffic flow velocity more accurately. However, while individuals are generally interested in the aggregate information provided by such a system, e.g., to estimate their commute time, they do not wish their own trajectories to be publicly revealed, since these might contain sensitive information about their driving behavior, frequently visited locations, etc. Privacy-preserving mechanisms for such location-based services are often based on ad-hoc temporal and spatial cloaking of the measurements (Gruteser and Grunwald 2003; Hoh et al. 2012). However, in the absence of a quantitative definition of privacy and a clear model of the adversary's capabilities, it is common that such techniques are later argued to be deficient (Shokri et al. 2009, 2010). The temporal cloaking scheme proposed in Hoh et al. (2012) for example aggregates the speed measurements of k users successively crossing a given line, but does not necessarily protect individual trajectories against adversaries exploiting temporal relationships between these aggregated measurements (Shokri et al. 2009).

The algorithms described in a following section provide ways to construct differentially private estimates of the signal (5.6). As discussed above, the adjacency relation can be placed on the measured (physical) input signals u_i, representing position measurements. If it is placed instead on the state signal x_i of the mathematical motion model, it must be kept in mind that this model is only an approximation, which assumes in particular that the acceleration of each car is a random white noise and that the parameters σ_1 and σ_2 are perfectly known, two assumptions that are unlikely to hold exactly in reality.

5.2.2 Collective Data Streams

A collective data stream $\{u_t\}_t \geq 0$, as defined in Sect. 2.3.1, is a partially aggregated signal, which could be directly captured by certain types of sensors. Again, we assume here the availability of a publicly known linear state-space model for the signal u, of the form

$$x_{t+1} = A_t x_t + B_t w_t \tag{5.7}$$
$$u_t = C_t x_t + D_t w_t, \tag{5.8}$$

where the matrices A_t, B_t, C_t, D_t are known, the noise w_t is zero-mean and white with identity covariance matrix, and the initial condition x_0 is a random variable with mean $\bar{x}_0$ and covariance $\bar{\Sigma}_0$. Our assumptions are identical to those presented for the model (5.1), (5.2), which now appears in fact as a special case of (5.7), (5.8), with the $A, B, C,$ and D matrices taken to be block-diagonal. The choice of adjacency relation on the input signals u could now be (2.5), (2.6) or (2.7), among other possibilities. As discussed in Sect. 5.2.1, it might also be reasonable for specific applications to define the adjacency relation on the state signal of the model (5.7), provided we can relate it to a physically meaningful signal that we want to protect, and with the same caveat that the privacy guarantee then becomes dependent on the confidence we have in the accuracy of our model.

In practice, a collective data stream u is an output of a "macroscopic" model of a given system, so that the size of the matrix A is significantly smaller than the sum of the dimensions of the individual state spaces obtained when we track each individual in (5.1)–(5.2). For example, the state x in (5.7) could represent the number of people (or the fraction of a given population) in N spatial regions (recall the building monitoring example of Sect. 3.5), in which case the model dimension scales with the number N of regions rather than the number of individuals. Since the state x for such macroscopic models represents an aggregate quantity instead of relating to a specific individual, it makes sense in this case to assume that the desired output of the filter is a full system state estimate $\hat{x}$.

5.3 Input Perturbation Mechanisms

The input perturbation mechanism of Sect. 2.3.2 can be combined with Kalman filtering in a straightforward manner. For personal data streams, each individual can add privacy-preserving white noise v_i directly on her signal u_i before sending it to the data aggregator. The level and distribution of the noise depends on the desired privacy guarantee. Suppose first that the adjacency relation is given by (2.3), defined directly on the input signal u, for some positive constant ρ. Following Corollary 2.1, for ε-differential privacy the norm used in (2.3) should be the ℓ_1-norm and each individual sends the signal $u_i + v_i$, with $v_{i,t} \sim \text{Lap}(\rho/\varepsilon)^{d_i}$, where d_i is the dimension of $u_{i,t}$. If only (ε, δ)-differential privacy is requested, with $\delta > 0$, then the ℓ_2-norm can be used in (2.3) and we can take $v_{i,t} \sim \mathcal{N}\left(0, \kappa_{\delta,\varepsilon}^2 \rho^2 I_{d_i}\right)$.

When the adjacency relation is defined on the state signal, as in (5.5), we can proceed as follows to set the noise level, e.g., for (ε, δ)-differential privacy. Since for two state trajectories x, x' adjacent according to (5.5), we have $x_i - x_i' = S_i(x_i - x_i')$, and moreover $S_i^2 = S_i$, we get the following sensitivity bound

$$\|u_i - u_i'\|_2 = \|C_i S_i (x_i - x_i')\|_2 = \|C_i S_i S_i (x_i - x_i')\|_2$$
$$\leq \sup_t \left\{\sigma_{\max}(C_{i,t} S_i)\right\} \rho.$$

Hence differential privacy can be guaranteed if each participant adds to its signal u_i a white Gaussian noise with covariance matrix $\kappa_{\delta,\varepsilon}^2 \rho^2 \sup_{i,t}\{\sigma_{\max}^2(C_{i,t} S_i)\} I_{d_i}$. Note that in this sensitivity computation the measurement noise $D_i w_i$ has the same realization independently of the considered variation in x_i.

Still for the case were the adjacency relation is defined on the state signal, as in (5.5), the properties of the Gaussian distribution also make it possible to leverage the natural measurement noise appearing in (5.2) to provide privacy. For simplicity, consider a time-invariant model. Assume that the noise $w_{i,t}$ in (5.1)–(5.2) is Gaussian, and that D_i is full row rank. Then, without further loss of generality, we can in fact assume that $D_i w_{i,t} = \bar{D}_i \bar{w}_{i,t}$, with $\bar{D}_i$ invertible and $\bar{w}_i$ a standard white Gaussian noise. Thus, we have

$$\bar{D}_i^{-1} u_i = \bar{D}_i^{-1} C_i x_{i,t} + \bar{w}_{i,t}. \tag{5.9}$$

As above, we have the sensitivity bound

$$\|\bar{D}_i^{-1} C_i (x_i - x_i')\|_2 \leq \Delta_i := \sigma_{\max}(\bar{D}_i^{-1} C_i S_i)\rho.$$

Let $\Delta = \max_i \Delta_i$. Hence, one can publish an (ε, δ)-differentially private version of the signals $\bar{D}_i^{-1} C_i x_{i,t}$ by adding Gaussian noise with variance $\kappa_{\delta,\varepsilon}^2 \Delta^2$ to them. Now, from (5.9), the signal $\bar{D}_i u_i$ already contains a Gaussian noise with unit variance. Since the Gaussian distribution is stable under linear combinations, we deduce that the signal $\bar{D}_i^{-1} u_i + v_i$, with v_i a white Gaussian noise independent of $\bar{w}_i$ and with covariance $\kappa_{\delta,\varepsilon}^2 \max\{\Delta^2 - 1, 0\} I_{d_i}$, is differentially private. By resilience to post-processing, we can premultiply this signal by $\bar{D}_i$ and thus publish the signal $u_i + \bar{D}_i v_i$, which is also differentially private. Again, as mentioned before, the main issue with this scheme is that the signal model is never perfect, and in particular the noise w_i might not be truly Gaussian for example, or its variance might not be perfectly known. Nevertheless, one could argue that the natural measurement noise in this case can help enforce some approximate version of differential privacy.

With input perturbation, the signals transmitted by the participants are already differentially private. Moreover, the level of privacy-preserving noise is known to the data aggregator, *which can then take this additional noise into account when designing the Kalman filter*. Indeed, the noise added to u_i above appears as an additional measurement noise. The final Kalman filter design with input perturbation is given in Algorithm 5.1, assuming the adjacency relation is placed directly on the input signals u_i to simplify the presentation. Note that these equations are valid for possibly correlated process and measurement noise, i.e., $B_{i,t} D_{i,t}^T \neq 0$ in the model (5.1)–(5.2). If Laplace noise is added to the signals u_i for ε-differential privacy, this filter corresponds to an LMMSE estimator. If Gaussian noise is added and the noise in the dynamics and measurements is also Gaussian, then the filter minimizes the MSE more generally among all possible filters.

Algorithm 5.1: Differentially Private Kalman Filter with Input Perturbation (case of individual data streams)

Data: A set of individual signals u_i, $1 \leq i \leq N$, modeled by (5.1), (5.2). Privacy parameters ε, δ and ρ.

Result: A differentially private signal $\hat{z}_t$ for the adjacency relation (2.3), where $p = 1$ or 2, with $\hat{z}_t$ an estimate of $z_t := \sum_{i=1}^n L_{i,t} x_{i,t}$ released at each period t.

Initialize: For all $1 \leq i \leq n$, let $\hat{x}_{i,0}^- = \bar{x}_{i,0}$, $\Sigma_{i,0}^- = \bar{\Sigma}_{i,0}$;

for *each* $t \geq 0$ **do**

> Perturb each signal sample $u_{i,t}$, $1 \leq i \leq n$, to $\tilde{u}_{i,t} = u_{i,t} + v_i$ with $v_{i,t} \sim \text{Lap}\left(\sqrt{2}\sigma\right)^{d_i}$, where $\sqrt{2}\sigma = \rho/\varepsilon$; or $v_{i,t} \sim \mathcal{N}(0, \sigma^2 I_{d_i})$, where $\sigma = \kappa_{\delta,\varepsilon}\rho$;
>
> Measurement update step: compute $K_{i,t} = \Sigma_{i,t}^- C_{i,t}^T (C_{i,t} \Sigma_{i,t}^- C_{i,t}^T + D_{i,t} D_{i,t}^T + \sigma^2 I)^{-1}$,
> $G_{i,t} = B_{i,t} D_{i,t}^T (D_{i,t} D_{i,t}^T + \sigma^2 I)^{-1}$, $x_{i,t}^+ = x_{i,t}^- + K_{i,t}(\tilde{u}_{i,t} - C_{i,t} x_{i,t}^-)$,
> $\Sigma_{i,t}^+ = (I - K_{i,t} C_{i,t})^T \Sigma_{i,t}^-$, for $1 \leq i \leq n$;
> Release $\hat{z}_t = \sum_{i=1}^n L_{i,t} \hat{x}_{i,t}^+$, and if desired the error covariance $P_t = \sum_{i=1}^n L_{i,t} \Sigma_{i,t}^+ L_{i,t}^T$;
> Prediction step: compute $\hat{x}_{i,t+1}^- = A_{i,t} \hat{x}_{i,t}^+ + G_{i,t}(\tilde{u}_{i,t} - C_{i,t} x_{i,t}^+)$,
> $\Sigma_{i,t+1}^- = (A_{i,t} - G_{i,t} C_{i,t}) \Sigma_{i,t}^+ (A_{i,t} - G_{i,t} C_{i,t})^T + B_{i,t} B_{i,t}^T - G_{i,t}(D_{i,t} D_{i,t}^T + \sigma^2 I) G_{i,t}^T$,
> for $1 \leq i \leq n$;

The design of Kalman filtering mechanisms with input perturbation for collective data streams, using the aggregate dynamic model (5.7)–(5.8), is essentially the same as for individual data streams. Namely, we add appropriate noise to the input signal u, and the Kalman filter subsequently takes this noise into account by treating it as additional measurement noise to estimate the state of the model (5.7). The only difference is this case is that we do not decompose the estimation problem into parallel subproblems any more in general.

5.4 Output Perturbation Mechanisms for Time-Invariant Filters

Following Theorem 2.3, a basic output perturbation mechanism consists in adding noise to the output of the Kalman filter, proportional to the filter's sensitivity, without changing its design. As always, the sensitivity calculation depends on the choice of adjacency relation, and on the norm associated to the chosen noise distribution. For the model (5.1)–(5.2), the equations of the Kalman filter are given by Algorithm 5.1 with $\sigma_i = 0$. Since we assume that $\bar{x}_{i,0}$ is public information, the initial condition $\hat{x}_{i,0}^{-}$ of each state estimator is fixed. If we want to add Gaussian noise on the filter estimate $\hat{z}_t = \sum_{i=1}^{n} L_{i,t} \hat{x}_{i,t}^{+}$ to achieve (ε, δ)-differential privacy, we need to compute the maximum ℓ_2-sensitivity of the n Kalman filters estimating the individual states from each individual input u_i. However, even if the adjacency relation is (2.3), obtaining a tight bound on this sensitivity is complicated by the fact that the Kalman filter is a time-varying system. Hence, for simplicity, we restrict our attention in this section to *time-invariant* filters, as well as time-invariant models (5.1)–(5.2). Recall that the time-invariant Kalman filter (Anderson and Moore 2005) uses at each period the gains K_i corresponding to the asymptotic limits $\lim_{t \to \infty} K_{i,t}$ of the gains in Algorithm 5.1. This filter is suboptimal in general for the performance measure (5.3) for finite T (unless $\bar{\Sigma}_{i,0}$ happens to be equal to the steady-state covariance matrix), but it is nonetheless often used in practice as it does not require recomputing (or storing) the gains and covariance matrices for each period. Moreover, it is optimal for the steady-state performance measure $e_{mse}^{KF, \infty}$. We have the following result.

Corollary 5.1 *Let $\varepsilon, \delta > 0$. Let $\mathscr{K}_i$ be the time-invariant Kalman filter for the system (5.1)–(5.2), for $1 \leq i \leq n$. A mechanism releasing $\left(\sum_{i=1}^{n} L_i \mathscr{K}_i u_i \right) + \nu$, with ν a white noise independent of $\{w_i\}_{1 \leq i \leq n}$, $\{x_{i,0}\}_{1 \leq i \leq n}$, is*

1. *(ε, δ)-differentially private for the adjacency relation (2.3), if $\nu_t \sim \mathscr{N}(0, \sigma^2 I)$ and $\sigma = \rho \kappa_{\delta, \varepsilon} \max_{1 \leq i \leq n} \{\|L_i \mathscr{K}_i\|_\infty\}$;*
2. *(ε, δ)-differentially private for the adjacency relation (5.5) if $\nu_t \sim \mathscr{N}(0, \sigma^2 I)$ and $\sigma = \rho \kappa_{\delta, \varepsilon} \max_{1 \leq i \leq n} \{\|L_i \mathscr{K}_i C_i S_i\|_\infty\}$.*

Proof This statement follows from Theorem 2.3 and Corollary 2.4. For the second item, suppose that the adjacency relation is (5.5). Consider two adjacent state trajectories x, x', and let $\hat{z}$, $\hat{z}'$ be the corresponding estimates produced by the time-invariant Kalman filter. With $\mathscr{K}_i$ denoting the time-invariant Kalman filter producing at each period the signal $\hat{x}_{i,t}^{+}$, we have

$$\hat{z} - \hat{z}' = L_i \mathscr{K}_i (u_i - u_i') = L_i \mathscr{K}_i C_i S_i (x_i - x_i')$$
$$= L_i \mathscr{K}_i C_i S_i S_i (x_i - x_i'),$$

using the fact that $S_i^2 = S_i$. Since the systems are now time-invariant, we have $\|\hat{z} - \hat{z}'\|_2 \leq \gamma_i \rho$, where γ_i is the H_∞ norm of the transfer function $L_i \mathscr{K}_i C_i S_i$.

5.4.1 Filter Redesign for Stable Systems

One can improve the output perturbation mechanism of Corollary 5.1 by redesigning the Kalman filter to optimize the overall (asymptotic) MSE performance. Indeed, this MSE depends on the estimation performance of the filter as well as the amount of privacy-preserving noise introduced at the output, which is a function of the H_∞ norm of the filter. Hence, we should design a filter that balances quality of estimation and size of its H_∞ norm. We consider therefore the design of n LTI filters of the form

$$\hat{x}_{i,t+1} = F_i \hat{x}_{i,t} + G_i u_{i,t} \tag{5.10}$$
$$\hat{z}_{i,t} = H_i \hat{x}_{i,t} + K_i u_{i,t}, \tag{5.11}$$

for $1 \le i \le n$, where F_i, G_i, H_i, K_i are matrices to determine. The overall system in front of the privacy-preserving noise source produces the estimate $\hat{z}_t = \sum_{i=1}^n \hat{z}_{i,t}$ of the signal z. We assume first in this section that the system matrices A_i of the model (5.1)–(5.2) are stable, in which case we also restrict the filter matrices F_i to be stable. Moreover, we only consider the design of full order filters, i.e., the dimensions of F_i are greater or equal to those of A_i, for all $1 \le i \le n$.

Denote the overall state for each system (5.1)–(5.2) and associated filter by $\tilde{x}_i = [x_i^T, \hat{x}_i^T]^T$. The combined dynamics from w_i to the estimation error $e_i := z_i - \hat{z}_i$ can be written

$$\tilde{x}_{i,t+1} = \tilde{A}_i \tilde{x}_{i,t} + \tilde{B}_i w_{i,t}$$
$$e_{i,t} = \tilde{C}_i \tilde{x}_{i,t} + \tilde{D}_i w_{i,t},$$

where

$$\tilde{A}_i = \begin{bmatrix} A_i & 0 \\ G_i C_i & F_i \end{bmatrix}, \quad \tilde{B}_i = \begin{bmatrix} B_i \\ G_i D_i \end{bmatrix}, \tag{5.12}$$
$$\tilde{C}_i = \begin{bmatrix} L_i - K_i C_i & -H_i \end{bmatrix}, \quad \tilde{D}_i = -K_i D_i.$$

The steady-state MSE for the ith estimator is then

$$\lim_{t \to \infty} \mathbb{E}[e_{i,t}^T e_{i,t}] = \|\tilde{C}_i (zI - \tilde{A}_i)^{-1} \tilde{B}_i + \tilde{D}_i\|_2^2.$$

We can now evaluate the impact on the overall MSE of the privacy-preserving noise added on the signals $\hat{z}_i$. For a change of a state trajectory x to an adjacent one x' according to (5.5), letting $\delta x_i = x_i - x_i' = S_i(x_i - x_i') = S_i \delta x_i$, we see that the change in the output of filter i follows the dynamics

$$\delta \hat{x}_{i,t+1} = F_i \delta \hat{x}_{i,t} + G_i C_i S_i \delta x_i$$
$$\delta \hat{z}_i = H_i \delta \hat{x}_{i,t} + K_i C_i S_i \delta x_i.$$

Hence, the ℓ_2-sensitivity can be measured by the H_∞ norm of the transfer function

$$\left[\begin{array}{c|c} F_i & G_i C_i S_i \\ \hline H_i & K_i C_i S_i \end{array} \right]. \tag{5.13}$$

Replacing the Kalman filter in Theorem 5.1, the MSE for the resulting output perturbation mechanism guaranteeing (ε, δ)-privacy is then

$$\left(\sum_{i=1}^{n} \| \tilde{C}_i (zI - \tilde{A}_i)^{-1} \tilde{B}_i + \tilde{D}_i \|_2^2 \right) + \kappa_{\delta,\varepsilon}^2 \rho^2 \max_{1 \leq i \leq n} \{ \gamma_i^2 \},$$

with

$$\gamma_i := \| H_i (zI - F_i)^{-1} G_i C_i S_i + K_i C_i S_i \|_{\infty}. \tag{5.14}$$

To minimize this MSE, we consider the following optimization problem

$$\min_{\mu_i, \lambda, F_i, G_i, H_i, K_i} \quad \sum_{i=1}^{n} \mu_i + \kappa_{\delta,\varepsilon}^2 \, \rho^2 \, \lambda \tag{5.15}$$

$$\text{s.t. } \forall \, 1 \leq i \leq n, \, \| \tilde{C}_i (zI - \tilde{A}_i)^{-1} \tilde{B}_i + \tilde{D}_i \|_2^2 \leq \mu_i, \tag{5.16}$$

$$\| H_i (zI - F_i)^{-1} G_i C_i S_i + K_i C_i S_i \|_{\infty}^2 \leq \lambda. \tag{5.17}$$

The following theorem gives a convex sufficient condition in the form of LMIs guaranteeing that a choice of filter matrices F_i, G_i, H_i, K_i satisfies the constraints (5.16)–(5.17). Its proof involves LMI techniques for the reformulation and optimization of control objectives (Scherer et al. 1997; Skelton et al. 1998) and can be found in Le Ny and Pappas (2014). Note that the $*$ notation represents symmetric terms in the matrices.

Theorem 5.1 *The constraints (5.16)–(5.17), for some $1 \leq i \leq n$, are satisfied if there exists matrices $W_i, Y_i, Z_i, \hat{F}_i, \hat{G}_i, \hat{H}_i, \hat{K}_i$ such that $Tr(W_i) < \mu_i$,*

$$\begin{bmatrix} W_i & (L_i - \hat{K}_i C_i - \hat{H}_i) & (L_i - \hat{K}_i C_i) & -\hat{K}_i D_i \\ * & Z_i & Z_i & 0 \\ * & * & Y_i & 0 \\ * & * & * & I \end{bmatrix} \succ 0,$$

$$\begin{bmatrix} Z_i & Z_i & Z_i A_i & Z_i A_i & Z_i B_i \\ * & Y_i & (M_i + \hat{F}_i) & M_i & (Y_i B_i + \hat{G}_i D_i) \\ * & * & Z_i & Z_i & 0 \\ * & * & * & Y_i & 0 \\ * & * & * & * & I \end{bmatrix} \succ 0, \text{ and } \begin{bmatrix} Z_i & Z_i & 0 & 0 & 0 & 0 \\ * & Y_i & 0 & \hat{F}_i & 0 & \hat{G}_i C_i S_i \\ * & * & \lambda I & \hat{H}_i & 0 & \hat{K}_i C_i S_i \\ * & * & * & Z_i & Z_i & 0 \\ * & * & * & * & Y_i & 0 \\ * & * & * & * & * & I \end{bmatrix} \succ 0,$$

where $M_i := Y_i A_i + \hat{G}_i C_i$. If these conditions are satisfied, one can recover admissible filter matrices F_i, G_i, H_i, K_i as

$$F_i = V_i^{-1} \hat{F}_i \hat{Z}_i^{-1} U_i^{-T}, \quad G_i = V_i^{-1} \hat{G}_i, \quad H_i = \hat{H}_i Z_i^{-1} U_i^{-T}, \quad K_i = \hat{K}_i \tag{5.18}$$

where U_i, V_i are any two nonsingular matrices such that $V_i U_i^T = I - Y_i Z_i^{-1}$.

Remark 5.1 Note that the problem (5.15) is also linear in μ_i, λ. These variables can then be minimized subject to the LMI constraints of Theorem 5.1 in order to design a good filter trading off estimation error and ℓ_2-sensitivity to minimize the overall MSE. However, including these variables directly in the optimization problem can lead to ill-conditioning in the inversion of the matrices U_i, V_i in (5.18), a phenomenon discussed together with a recommended fix in Scherer et al. (1997, p. 903). In addition, minimizing the objective (5.15) subject to the LMI constraints is different from solving (5.15)–(5.17), due to the conservativeness of the conditions in Theorem 5.1. As in mixed H_2/H_∞ problems, one could consider more complex algorithms to reduce this conservativeness (Scherer 2000). Consider now, instead of (5.15), the objective $\sum_{i=1}^{n} \mu_i + \alpha \lambda$, where the parameter $\alpha \geq 0$ replaced $\kappa_{\delta,\varepsilon}^2 \rho^2$, subject to the LMIs of Theorem 5.1. By setting $\alpha = 0$, we recover exactly the Kalman filter. Hence by performing a one-dimensional search over α we can attempt to improve the overall MSE of the output mechanism over the basic Kalman filter design.

5.4.2 Unstable Systems

If the dynamics (5.1) are not stable, the design approach presented in Sect. 5.4.1 needs to be modified by further restricting the class of filters considered. As before we minimize the estimation error variance together with the sensitivity measured by the H_∞ norm of the filter. Starting from the filter dynamics (5.10), (5.11), we can consider designs where $\hat{x}_i$ is an estimate of x_i, and set $H_i = L_i$, $K_i = 0$, so that $\hat{z}_i = L_i \hat{x}_i$ is an estimate of $z_i = L_i x_i$. The error dynamics $e_i := x_i - \hat{x}_i$ then satisfies

$$e_{i,t+1} = (A_i - G_i C_i)x_{i,t} - F_i \hat{x}_{i,t} + (B_i - G_i D_i)w_{i,t}.$$

Setting $F_i = (A_i - G_i C_i)$ gives an error dynamics independent of x_i

$$e_{i,t+1} = (A_i - G_i C_i)e_{i,t} + (B_i - G_i D_i)w_{i,t}, \tag{5.19}$$

and leaves the matrix G_i as the only remaining design variable. Note however that the resulting class of filters contains the (one-step delayed) Kalman filter. To obtain a bounded error, there is an implicit constraint on G_i that $A_i - G_i C_i$ should be stable.

Now, following the discussion in the previous subsection, minimizing the MSE while enforcing differential privacy leads to the following optimization problem

$$\min_{\mu_i, \lambda, G_i} \quad \sum_{i=1}^{n} \mu_i + \kappa_{\delta,\varepsilon}^2 \rho^2 \lambda \tag{5.20}$$

$$\text{s.t. } \|L_i(zI - (A_i - G_i C_i))^{-1}(B_i - G_i D_i)\|_2^2 \leq \mu_i, \tag{5.21}$$

$$\|L_i(zI - (A_i - G_i C_i))^{-1}G_i C_i S_i\|_\infty^2 \leq \lambda, \ \forall\ 1 \leq i \leq n. \tag{5.22}$$

Again, the following theorem, whose proof can be found in Le Ny and Pappas (2014), provides a sufficient condition taking the form of LMIs and guaranteeing that the constraints (5.21)–(5.22) are satisfied. Optimizing over the variables λ, μ_i, G_i can then be done using semidefinite programming.

Theorem 5.2 *The constraints (5.21)–(5.22), for some $1 \leq i \leq n$, are satisfied if there exists matrices Y_i, X_i, $\hat{G}_i$ such that*

$$Tr(Y_i L_i^T L_i) < \mu_i, \quad \begin{bmatrix} Y_i & I \\ I & X_i \end{bmatrix} \succ 0,$$

$$\begin{bmatrix} X_i & X_i A_i - \hat{G}_i C_i & X_i B_i - \hat{G}_i D_i \\ * & X_i & 0 \\ * & * & I \end{bmatrix} \succ 0, \tag{5.23}$$

$$and \quad \begin{bmatrix} X_i & 0 & X_i A_i - \hat{G}_i C_i & \hat{G}_i C_i S_i \\ * & \lambda I & L_i & 0 \\ * & * & X_i & 0 \\ * & * & * & I \end{bmatrix} \succ 0. \tag{5.24}$$

If these conditions are satisfied, one can recover an admissible filter matrix G_i by setting $G_i = X_i^{-1} \hat{G}_i$.

5.4.3   Post-filtering

With the output perturbation mechanisms presented above, we have the undesirable situation that the output signal is directly contaminated by the privacy-preserving noise. From an estimation performance point-of-view, this is clearly suboptimal and this output noise should again be filtered. Now, since the filters discussed above in this section are themselves linear systems, and since the privacy-preserving noise is zero-mean with known variance, we can in fact add another Kalman filter to realize this post-filtering. This Kalman filter at the output does not need to be time-invariant, since we use it only for post-processing and hence do not need to compute its sensitivity. Its design depends both on the parameters of the model (5.1)–(5.2) and of those of the first-stage filter, which compose together in a cascade. The first-stage filter is either a time-invariant Kalman filter as in Corollary 5.1, or a redesigned filter as presented in Sects. 5.4.1 and 5.4.2. This construction is depicted on Fig. 5.1.

On Fig. 5.1, the prefilter is now the filter designed in Sect. 5.4.1 or 5.4.2. Given the model (5.1)–(5.2) for the individual signal dynamics, and (5.10)–(5.11) for the prefilter, we have that the signal y on Fig. 5.1 is generated by the following LTI system

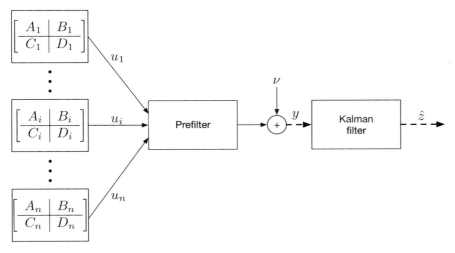

Fig. 5.1 Adding a post-filtering stage to the output perturbation mechanism. The prefilter could be a time-invariant Kalman filter as in Corollary 5.1 or a redesigned LTI filter as presented in Sects. 5.4.1 and 5.4.2. Yet a better design is to follow the methodology of Chap. 3 and optimize the prefilter with the knowledge that it is followed by a Kalman filter for post-processing, as discussed in Sect. 5.5

$$\tilde{x}_{i,t+1} = \tilde{A}_i \tilde{x}_{i,t} + \tilde{B}_i w_{i,t}, \quad 1 \le i \le n,$$

$$y_t = \sum_{i=1}^{n} \left[K_i C_i \ H_i \right] \tilde{x}_{i,t} + \sum_{i=1}^{n} K_i D_i w_{i,t} + v_t$$

where the matrices $\tilde{A}_i$, $\tilde{B}_i$ are defined in (5.12) and v is white zero-mean Gaussian noise with covariance matrix $\sigma_v^2 I$, where $\kappa_{\delta,\varepsilon}^2 \rho^2 \max_{1 \le i \le n}\{\gamma_i^2\}$ and γ_i defined in (5.14). Since this dynamical system has states $\tilde{x}_i = [x_i^T, \hat{x}_i^T]^T$, with $\hat{x}_i$ the internal state of the prefilter i, we are again interested in publishing at the output of the second Kalman filter an estimate of $L_i x_{i,t}$, i.e., we discard the estimate of the internal state of the prefilter.

To close this section, we can now remark that the architecture of Fig. 5.1 is again a two-stage architecture, as introduced in Chap. 3, but we have followed here a suboptimal approach to design it. Indeed, we designed the prefilter first by assuming an output perturbation mechanism, i.e., not taking into account the possibility of post-filtering the privacy-preserving noise. We then "cleaned" the output y of the output perturbation mechanism to smooth out the output noise v, which can be done using a Kalman filter since all the dynamic systems preceding y are linear. A better approach outlined in Chap. 3 is to optimize the prefilter given the knowledge that the second stage is a Kalman filter, a choice that is justified as long as the prefilter is linear, and is in fact optimal when all noise sources are Gaussian, including the privacy-preserving noise. Section 5.5 discusses this first stage optimization approach for differentially private Kalman filtering, although only in the simplified case of static prefilters.

5.4.4 Example

Let us illustrate the behavior of the differentially private Kalman filters introduced so far in the context of the average speed estimation problem of Example 5.1. All individual systems are identical, hence we drop the subscript i in the notation. Assume that the selection matrix is $S = \begin{bmatrix} 1 & 0 \\ 0 & 0 \end{bmatrix}$, that $\rho = 100$ m in (5.5), and that we have $n = 200$ participants. Let $T_s = 1$ s, $\sigma_1 = 1\,\text{m} \cdot \text{s}^{-2}$, $\sigma_2 = 10\,\text{m}$. A single Kalman filter denoted $\mathscr{K}$ is designed to provide an estimate $\hat{x}_i$ of each state vector x_i, so that in absence of privacy constraint the final estimate would be

$$\hat{z} = \begin{bmatrix} 0 & \frac{1}{n} \end{bmatrix} \sum_{i=1}^{n} \mathscr{K}\, y_i = \begin{bmatrix} 0 & 1 \end{bmatrix} \mathscr{K} \left(\frac{1}{n} \sum_{i=1}^{n} y_i \right).$$

We designed the Kalman filters for various values of the privacy parameters ε, δ. For the output perturbation mechanisms, we used the approach described in Remark 5.1 to trade-off estimation error and $\mathscr{H}_\infty$-norm of the filter. However, in this case this resulted in only a marginal improvement of the MSE of the Kalman filter based output perturbation mechanism. Hence, for conciseness we restrict the following discussion of output perturbation mechanisms to the simplest scheme that does not redesign the original Kalman filter. In particular, for the two-stage mechanism of Fig. 5.1, the prefilter is also a (time-invariant) Kalman filter.

Figure 5.2 shows the steady-state Root-Mean-Square Error (RMSE) of the mechanisms for different values of ε, with fixed $\delta = 0.05$. The input perturbation mechanism is essentially unusable with the original Kalman filter, but shows reasonably good performance when the filter is redesigned by taking the privacy-preserving noise into account as additional measurement noise, especially in the high privacy regime (small ε). However, the best RMSE performance among these four different designs is obtained with the two-stage mechanism using two Kalman filters.

Other measures of performance than the RMSE are also of interest, and in particular on Fig. 5.3 one can see the difference in convergence time for the various mechanisms. Here, the filters are simply initialized with an incorrect value of the initial average velocity, but this also serves to illustrate situations where we could have a sudden change in traffic velocity, e.g., due to the formation of a traffic jam. In such cases, it is desirable to have fast convergence of the filter, e.g., in order to warn downstream drivers sufficiently soon. For $\varepsilon = 0.3$, $\delta = 0.05$, the output perturbation mechanism converges in few seconds, whereas the input perturbation mechanism takes more than a minute to converge. In this case, the slightly higher asymptotic RMSE of the first mechanism, which nonetheless remains below 2 km/h, seems to be acceptable in view of the much improved convergence speed. The two-stage mechanism has a slightly higher convergence time than the output mechanism due to the delay introduced by the second stage, but overall still provide a good trade-off between accuracy and speed. As ε, δ decrease and hence the privacy-preserving noise increases, the convergence speed of the input perturbation mechanism degrades

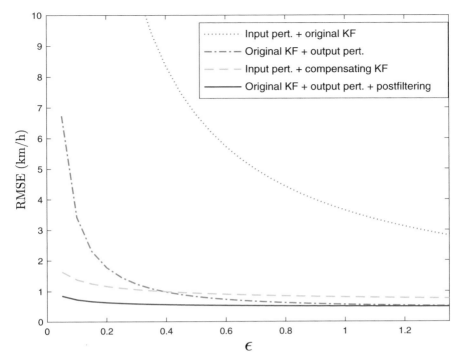

Fig. 5.2 Steady-state RMSE of the average velocity estimate for three mechanisms, as a function of the privacy parameter ε

quickly (Le Ny and Pappas 2014). To a lesser extent, this might also be the case for the two-stage mechanism, and so as a result, there might be situations in the high privacy regime where it is beneficial to leave the noise of the output perturbation mechanism unfiltered.

5.5 Optimization of Two-Stage Kalman Filtering Mechanisms: Static First Stage Aggregation

Considering Fig. 5.1 again, we can interpret a two-stage mechanism as combining the privacy-sensitive signals in an appropriate way before adding the privacy preserving noise. To motivate the discussion of this section, let us show with a simple example that even just taking linear combinations, i.e., choosing the prefilter to be linear and memoryless, can be sufficient to improve significantly the performance over the input perturbation mechanism. Namely, consider the scalar case of model (5.1)–(5.2) with $A_{i,t} = a$, $C_{i,t} = c$,

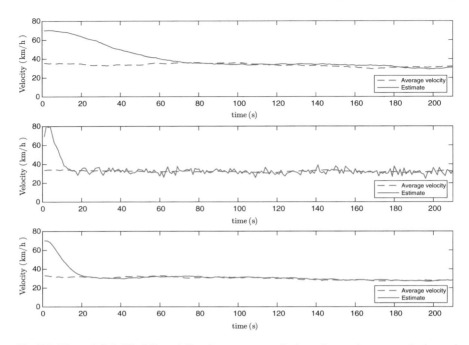

Fig. 5.3 Three $(0.3, 0.05)$-differentially private average velocity estimates: input perturbation and compensating Kalman filter (top), output perturbation and original Kalman filter (middle), adding a second Kalman filter to smooth out the noise of the output perturbation scheme (bottom). The filters are initialized with the same incorrect initial mean velocity (70 km/h instead of 35 km/h). The input perturbation mechanism shows good asymptotic accuracy but bad convergence time, whereas the output perturbation scheme shows the opposite. Using a cascade of two Kalman filters around the privacy-preserving noise gives the best asymptotic accuracy and reasonably good transient performance at this privacy level

$$\begin{bmatrix} B_{i,t} \\ D_{i,t} \end{bmatrix} = \begin{bmatrix} \sigma_w & 0 \\ 0 & \sigma_v \end{bmatrix}$$

and $L_{i,t} = 1$, so that we wish to estimate $z_t = \sum_{i=1}^{n} x_{i,t}$. The adjacency relation is (2.3) for a given value of ρ and the performance measure for an estimate $\hat{z}$ of z is the steady-state MSE $e_{mse}^{KF,\infty}$, see (5.4). The input perturbation mechanism with Gaussian noise corresponds to collecting the signal u_t with dynamics

$$\begin{aligned} x_{t+1} &= A_t x_t + \sigma_w w_t, \\ u_t &= C_t x_t + \sigma_v v_t + \zeta_t, \end{aligned} \tag{5.25}$$

with $A_t = a I_n$, $C_t = c I_n$, $w_t = \begin{bmatrix} w_{1,t} \dots w_{n,t} \end{bmatrix}^T$ and $v_t = \begin{bmatrix} v_{1,t} \dots v_{n,t} \end{bmatrix}^T$ standard white noises, and ζ_t the privacy-preserving white noise with distribution $\zeta_t \sim \mathcal{N}\left(0, \sigma_{\text{priv}}^2 I_n\right)$, where $\sigma_{\text{priv}} = \kappa_{\delta,\varepsilon} \rho$, see Algorithm 5.1. Given the symmetry of the problem, the steady-state MSE achievable by Kalman filtering can then be computed in a straightforward way by solving an algebraic Riccati equation (ARE), to obtain (see Degue and Le Ny 2017b for details)

$$MSE_1 = \frac{n}{2c^2} \left(-\beta + \sqrt{\beta^2 + 4c^2(\sigma_v^2 + \sigma_{\text{priv}}^2)\sigma_w^2} \right), \tag{5.26}$$

where $\beta = (1 - a^2)(\sigma_v^2 + \sigma_{\text{priv}}^2) - c^2\sigma_w^2$. We see in particular that the MSE per input signal MSE_1/n includes a constant penalty due to the presence of σ_{priv} (which also influences β), independently of the number of signals n.

Instead of using input perturbation, we can use the architecture of Fig. 5.1, where we now assume that the prefilter is simply the constant matrix $G = \mathbf{1}_n^{\mathsf{T}}$, i.e., a $1 \times n$ row vector of ones. Consider the same adjacency relation and denote $h_t = Du_t = \sum_{i=1}^n u_{i,t}$, $\xi_t = \sum_{i=1}^n w_{i,t}$ and $\eta_t = \sum_{i=1}^n v_{i,t}$, where w and v are defined in (5.25). We have

$$z_{t+1} = az_t + \sigma_w \xi_t,$$
$$h_t = cz_t + \sigma_v \eta_t.$$

Since

$$\sup_{u,u':\text{Adj}(u,u')} |Gu - Gu'| = \max_{1 \le i \le n} |u_i - u_i'| \le \rho,$$

releasing $y_t = h_t + v_t$, with v Gaussian white noise such that $v_t \sim \mathcal{N}(0, \sigma_{\text{priv}}^2)$, for the same value $\sigma_{\text{priv}} = \kappa_{\delta,\varepsilon}\rho$, is (ε, δ)-differentially private for the adjacency relation (2.3). Again, one can then design a steady-state Kalman filter taking this scalar signal y_t as input to estimate z_t. The steady-state MSE for this estimator is obtained by solving another scalar ARE, which gives (again, see Degue and Le Ny 2017b for details)

$$MSE_2 = \frac{n}{2c^2} \left(-\beta_{(n)} + \sqrt{\beta_{(n)}^2 + 4c^2 \left(\sigma_v^2 + \frac{\sigma_{\text{priv}}^2}{n} \right) \sigma_w^2} \right). \tag{5.27}$$

Comparing (5.26) and (5.27), we see that the only difference is the vanishing influence of the privacy preserving noise on the performance per signal MSE_2/n as n increases, as the term σ_{priv} in MSE_1 and β is replaced by $\sigma_{\text{priv}}/\sqrt{n}$ in MSE_2 and $\beta_{(n)}$. In particular, with the first-stage aggregation, the MSE per signal MSE_2/n improves with n and in fact converges to the performance of the non private Kalman filter as n increases. It is indeed a desirable feature of a good privacy-preserving mechanism that it should be easier to enforce privacy in publishing an aggregate statistic when the number of individuals in the database increases. Unfortunately, this is not a property exhibited by the input perturbation mechanism, but we see in this case that even a simple static pre-aggregation is sufficient to recover a much better asymptotic behavior.

The previous example and previous chapters motivate the problem of optimizing the two-stage differentially private Kalman filtering architecture of Fig. 5.1. We only consider here a restricted version of this problem, assuming the first stage to be simply a matrix G rather than a dynamic filter. Hence, consider the n individual dynamics (5.1)–(5.2). We make the simplifying assumptions that the process and measurement noises are uncorrelated and that the measurement noise covariance is time-invariant. That is, $D_{i,t} = D_i$ and $B_{i,t} D_i^T = 0$, for all $1 \le i \le n$ and all t, and we define the notation $W_{i,t} := B_{i,t} B_{i,t}^T$, $V_i := D_i D_i^T$ for the process and measurement noise covariance matrices respectively, for $1 \le i \le n$. Moreover we make the assumptions $W_{i,t} \succ 0$ and $V_i \succ 0$ for all t and i. The global system dynamics can be expressed as (5.7)–(5.8) with the block diagonal matrices $A_t := \text{diag}(A_{1,t}, \ldots, A_{n,t})$, $C_t := \text{diag}(C_{1,t}, \ldots, C_{n,t})$, $B_t B_t^T = W_t := \text{diag}(W_{1,t}, \ldots, W_{n,t})$, $D D^T = V := \text{diag}(V_1, \ldots, V_n)$, and $B_t D^T = 0$. Also, let $L_t := \begin{bmatrix} L_{1,t} & \ldots & L_{n,t} \end{bmatrix}$, so that $z_t = L_t x_t = \sum_{i=1}^{n} L_{i,t} x_{i,t}$ is the quantity we wish to estimate.

Following Fig. 5.1, the global signal u is first transformed to a signal Gu, where $G := \begin{bmatrix} G_1 & \ldots & G_n \end{bmatrix}$ is a matrix to determine, with a decomposition compatible with that of the signal u, i.e., $G_i \in \mathbb{R}^{p \times d_i}$ and the number p of rows of G is left to determine. The transformed signal Gu is then perturbed by additive white noise ν to obtain a differentially private signal. Let us use here the Gaussian mechanism. By Corollary 2.4, the signal $y_t = Gu_t + \nu_t$, with $\nu_t \sim \mathcal{N}(0, (\kappa_{\delta,\varepsilon} \Delta_2 G)^2 I_p)$, is (ε, δ)-differentially private for the adjacency relation (2.3), where the ℓ_2 sensitivity $\Delta_2 G$ of G is

$$\Delta_2 G = \sup_{\text{Adj}(u,u')} \|Gu - Gu'\|_2 = \max_i \left\{ \sup_{\|u_i - u_i'\|_2 \le \rho} \|G_i u_i - G_i u_i'\|_2 \right\}$$

$$\Delta_2 G = \rho \max_i \{\sigma_{max}(G_i)\}. \tag{5.28}$$

The state-space model for the signal y is now

$$x_{t+1} = A_t x_t + \xi_t \tag{5.29}$$

$$y_t = GC x_t + \zeta_t, \tag{5.30}$$

where $\xi_t = B_t w_t$ is a white noise with covariance W_t, and $\zeta_t = G D_t w_t + \nu_t$ is a white noise uncorrelated with ξ and with covariance $G V_t G^T + (\kappa_{\delta,\varepsilon} \Delta_2 G)^2 I_p$. Since this model is linear, we can design a Kalman filter (second stage on Fig. 5.1) to estimate z from the signal y. The equations for this Kalman filter depend on G, which enters the measurement Eq. (5.30) and the covariance of the measurement noise. Now, recalling Step 3 of our design methodology presented in Sect. 3.2, once the structure of the post-filter is fixed, we should express the performance measure as a function of G. Since $z = L_t x_t$ and $\hat{z}_t = L_t \hat{x}_t^+$, where $\hat{x}^+$ is the state estimate obtained right after a measurement update, we have

$$\mathbb{E}[(z_t - \hat{z}_t)(z_t - \hat{z}_t)^T] = L_t \Sigma_t L_t^T,$$

$$\text{and } e_{mse}^{KF,T} = \frac{1}{T} \sum_{t=0}^{T-1} \mathbb{E}\left[\|z_t - \hat{z}_t\|_2^2\right] = \frac{1}{T} \sum_{t=0}^{T-1} \text{Tr}(L_t \Sigma_t L_t^T), \tag{5.31}$$

where $\Sigma_t = \mathbb{E}[(x_t - \hat{x}_t^+)(x_t - \hat{x}_t^+)^T | y_{0:t}]$, $0 \leq t \leq T$, are the error covariance matrices. For the Kalman filter, these matrices follow the Riccati iterations written here in the information filter form

$$\begin{aligned}
\Sigma_0^{-1} &= (\bar{\Sigma}_0)^{-1} + C_0^T \Pi C_0, \\
\Sigma_{t+1}^{-1} &= (A_t \Sigma_t A_t^T + W_t)^{-1} + C_{t+1}^T \Pi C_{t+1}, \quad 0 \leq t \leq T - 1,
\end{aligned} \tag{5.32}$$

with $\bar{\Sigma}_0 = \text{diag}(\bar{\Sigma}_{1,0}, \ldots, \bar{\Sigma}_{n,0})$ the initial covariance matrix and Π is a $d \times d$ matrix that depends on G as follows

$$\Pi := G^T (GVG^T + (\kappa_{\delta,\varepsilon} \, \Delta_2 G)^2 I_p)^{-1} G. \tag{5.33}$$

We now see that the optimal static pre-filter G is a matrix that minimizes the objective (5.31) subject to the constraints (5.32), where G appears in the expression of Π given in (5.33). Once G is fixed, so is the sequence of covariance matrices as well as the sequence of gain matrices for the Kalman filter. We have thus recast our two-stage filtering problem as an optimization problem over the static pre-filters G.

As in Sect. 5.4, it turns out that the optimization of the G matrix can be reformulated as a semidefinite program (SDP), i.e., a convex optimization problem. The SDP can then be passed to standard solvers to obtain a solution guaranteed to be optimal, although computational issues remain due the large size of the SDP to solve. We skip here the presentation of the detailed discussion for this reformulation and refer the reader instead to Degue and Le Ny (2017a, b) for the missing technical details. Overall, one can transform the initial optimization problem for G to the following SDP with positive semidefinite matrix variables Π, X_t and Ω_t, for all $0 \leq t \leq T$

$$\min_{\{X_t, \Omega_t\}_{0 \leq t \leq T}, \Pi \succeq 0} \frac{1}{T+1} \sum_{t=0}^{T} \text{Tr}(X_t) \tag{5.34a}$$

$$\text{s.t.} \quad \begin{bmatrix} X_t & L_t \\ L_t^T & \Omega_t \end{bmatrix} \succeq 0, \quad 0 \leq t \leq T, \tag{5.34b}$$

$$\Omega_0 = \bar{\Sigma}_0^{-1} + C_0^T \Pi C_0, \tag{5.34c}$$

$$\begin{bmatrix} C_{t+1}^T \Pi C_{t+1} - \Omega_{t+1} + W_t^{-1} & W_t^{-1} A_t \\ A_t^T W_t^{-1} & \Omega_t + A_t^T W_t^{-1} A_t \end{bmatrix} \succeq 0, \quad 0 \leq t \leq T - 1, \tag{5.34d}$$

$$\begin{bmatrix} \frac{1}{\kappa_{\delta,\varepsilon}^2 \rho^2} I_{d_i} + V_i^{-1} & E_i^T \\ E_i & V - V \Pi V \end{bmatrix} \succeq 0, \quad 1 \leq i \leq n. \tag{5.34e}$$

In (5.34e), we introduce the $d \times d_i$ matrices $E_i = \begin{bmatrix} 0 & \dots & I_{d_i} & \dots & 0 \end{bmatrix}^T$, for $1 \leq i \leq n$, whose elements are zero except for an identity matrix in the ith block. These n LMIs essentially capture the sensitivity constraint, whereas the other LMIs capture the Riccati iteration constraints. The matrix variables Ω_t are related to, but not exactly the same as the inverse covariance matrices Σ_t in (5.32), see Degue and Le Ny (2017b) for details. We have the following theorem (Degue and Le Ny 2017b).

Theorem 5.3 *Let $\Pi^* \succeq 0, \{X_t^* \succeq 0, \Omega_t^* \succ 0\}_{0 \leq t \leq T}$ be an optimal solution for (5.34a)–(5.34e). Suppose that for some $0 \leq t \leq T$, we have $L_t(\Omega_t^*)^{-1} C_t^T \neq 0$. Then there exists a matrix G^* satisfying (5.33) minimizing the objective (5.31) for the two-stage Kalman filtering architecture with static linear prefilter, and the corresponding optimum MSE is equal to the optimum value of the SDP. This matrix G also satisfies $\rho \sigma_{\max}(G_i^*) = 1$ for all $1 \leq i \leq n$.*

To compute the optimum matrix G^*, we start by solving the SDP to obtain in particular an optimal $d \times d$ matrix Π^*. We then perform an eigenvalue or singular value decomposition of the following matrix, omitting the zero eigenvalues,

$$\kappa_{\delta,\varepsilon}^2 \left[(V - V\Pi^* V)^{-1} - V^{-1} \right] = Q \Lambda Q^T,$$

where $Q \in \mathbb{R}^{d \times p}$ has p orthonormal columns and Λ is a $p \times p$ diagonal matrix with positive entries. We then set $G^* = \Lambda^{1/2} Q^T$. By construction, this matrix is such that $\Delta_2 G^* = 1$.

Remark 5.2 The condition $L_t(\Omega_t^*)^{-1} C_t^T \neq 0$ for some t appears to be a weak requirement to guarantee the possibility of reconstructing the matrix G, required for technical reasons in the proof of Theorem 5.3.

In the stationary case ($T \to \infty$), with the dynamics now assumed time-invariant, a matrix G preceding a time-invariant Kalman filter can be computed by solving the following smaller SDP with symmetric matrix variables $\Pi \succeq 0$, $X \succeq 0$, $\Omega \succ 0$

$$\min_{X, \Omega, \Pi} \quad \mathrm{Tr}(X) \tag{5.35a}$$

$$\text{s.t.} \quad \begin{bmatrix} X & L \\ L^T & \Omega \end{bmatrix} \succeq 0, \tag{5.35b}$$

$$\begin{bmatrix} C^T \Pi C - \Omega + \Xi & \Xi A \\ A^T \Xi & \Omega + A^T \Xi A \end{bmatrix} \succeq 0, \tag{5.35c}$$

$$\begin{bmatrix} \frac{1}{\kappa_{\delta,\varepsilon}^2 \rho^2} I_{d_i} + V_i^{-1} & E_i^T \\ E_i & V - V\Pi V \end{bmatrix} \succeq 0, \quad 1 \leq i \leq n. \tag{5.35d}$$

The optimal matrix G is obtained by matrix factorization from the optimal Π^* as explained below Theorem 5.3. The steady-state MSE $e_{mse}^{KF,\infty}$ of (5.4) is equal to the optimum value of the SDP (5.35a)–(5.35d). Solving this stationary problem is clearly easier than for the previous time-varying case, due to the much reduced number of variables in the SDP.

References

Anderson BDO, Moore JB (2005) Optimal filtering. Dover, New York

Degue KH, Le Ny J (2017a) On differentially private Kalman filtering. In: Proceedings of the IEEE global conference on signal and information processing (GlobalSIP), Montreal, Canada

Degue KH, Le Ny J (2017b) Two-stage architecture optimization for differentially private Kalman filtering. Technical report, Polytechnique, Montreal. https://arxiv.org/abs/1707.08919

Gruteser M, Grunwald D (2003) Anonymous usage of location-based services through spatial and temporal cloaking. In: Proceedings of the 1st international conference on mobile systems, applications and services (MobiSys' 03), pp 31–42

Hoh B et al (2012) Enhancing privacy and accuracy in probe vehicle based traffic monitoring via virtual trip lines. IEEE Trans Mob Comput 11:5

Le Ny J, Pappas GJ (2014) Differentially private filtering. IEEE Trans Autom Control 59(2):341–354

Scherer CW (2000) An efficient solution to multi-objective control problems with LMI objectives. Syst Control Lett 40:43–57

Scherer C, Gahinet P, Chilali M (1997) Multiobjective output-feedback control via LMI optimization. IEEE Trans Autom Control 42(7):896–911

Shokri R et al (2009) A distortion-based metric for location privacy. In: Proceedings of the CCS workshop on privacy in the electronic society (WPES)

Shokri R et al (2010) Unraveling an old cloak: k-anonymity for location privacy. In: Proceedings of the CCS workshop on privacy in the electronic society (WPES)

Skelton RE, Iwasaki T, Grigoriadis K (1998) A unified algebraic approach to linear control design. Taylor and Francis, Routledge

Chapter 6
Differentially Private Nonlinear Observers

6.1 Introduction

[1]In the previous chapters, our focus has been on linear estimation and filtering problems, motivated by the fact that linear signal processing is both very useful in practice and supported by efficient design methods, which lead to tight trade-offs between accuracy and privacy. Nonetheless, in many cases the phenomena and systems that are of interest when processing privacy-sensitive signals obtained from a population of individuals have inherently nonlinear dynamics. As a result, accurate estimation of these phenomena might also require nonlinear estimators, e.g., in the form of models reproducing the nonlinear dynamics to perform accurate predictions. Consider for example the problem of detecting a disease outbreak in a population based on health data such as reported counts of infected individuals by health centers over time, a problem that we discuss in more details in Sect. 6.6. A model-based estimator and predictor for the proportion of infected individuals in the population might leverage an epidemiological model for the disease, but such models must typically be nonlinear to capture the possibility of multiple equilibria (e.g., the disease-free equilibrium and an equilibrium where the disease remains present), or periodic resurgence of the disease.

As we saw already in Chap. 2, a basic method to achieve differential privacy for a publicly released signal is to add noise proportional to the sensitivity of the dynamical system processing the privacy-sensitive data to produce this signal. Moreover, this sensitivity can be interpreted as a form of incremental system gain, at least under some choice of adjacency relation defining the differential privacy guarantee offered. Bounding the sensitivity is also a building block to develop more complex architectures, such as the two-stage architecture introduced in Chap. 3. For nonlinear systems, various tools could be used to bound the system sensitivity, depending on

---

[1]Some of the text in the Sects. 6.2–6.6 of this chapter is reproduced from Le Ny (2018) by permission of John Wiley & Sons, Ltd.

J. Le Ny, *Differential Privacy for Dynamic Data*,
SpringerBriefs in Control, Automation and Robotics,
https://doi.org/10.1007/978-3-030-41039-1_6

the characteristics of the specific application considered. In this chapter, our discussion focuses on systems for which one can establish a certain contraction property (Lohmiller and Slotine 1998; Sontag and Willems 2010; Forni and Sepulchre 2014; Angeli 2000).

The rest of this chapter, which is based on the results presented in Le Ny (2015, 2018), is divided as follows. Section 6.2 presents the problem statement formally, and discusses issues specific to nonlinear systems when comparing input and output perturbation mechanisms. In Sect. 6.4 we discuss some fundamental results in contraction analysis and present a type of "input-to-state stability" property of contracting systems. This property is used in Sect. 6.5 to design differentially private observers with output perturbation. The methodology is illustrated via two examples involving the analysis of dynamic data originating from private individuals. In Sect. 6.6, we consider first the problem of estimating link formation probabilities in a social network using a dynamic version of the classical stochastic block model (Holland et al. 1983), which involves a nonlinear measurement model. Then, we consider a nonlinear epidemiological model and design a differentially private estimator of the proportion of susceptible and infectious people in a population, based on syndromic data.

The following notation and definitions will be used in this chapter. The set of continuously differentiable functions is denoted $\mathscr{C}^1$. A class $\mathscr{K}$ function $\beta : \mathbb{R}_+ \to \mathbb{R}_+$ is a strictly increasing continuous function such that $\beta(0) = 0$. A continuous function $\beta : \mathbb{R}_+ \times \mathbb{R}_+ \to \mathbb{R}_+$ is of class $\mathscr{K}\mathscr{L}$ if $\beta(\cdot, s)$ is of class $\mathscr{K}$ for each value of s, and for each r, the function $s \mapsto \beta(r, s)$ is decreasing and tends to 0 as $s \to \infty$. For $H : \mathsf{X} \to \mathsf{Y}$ a linear map between finite dimensional vector spaces X and Y equipped with the norms $|\cdot|_\mathsf{X}$ and $|\cdot|_\mathsf{Y}$ respectively, we denote by $\|H\|_\mathsf{Y}^\mathsf{X}$ its induced norm, so that $|Hx|_\mathsf{Y} \leq \|H\|_\mathsf{Y}^\mathsf{X} |x|_\mathsf{X}$, for all x in X. If $\mathsf{X} = \mathsf{Y}$ and both spaces are equipped with the same norm $|\cdot|_\mathsf{X}$, we simply write $\|\cdot\|_\mathsf{X}$. We use $\mathrm{diag}(v)$ to denote a diagonal matrix with the components of the vector v on the diagonal. For $P \succeq 0$, we denote its (unique) positive semi-definite square root as $P^{1/2}$, i.e., $P = P^{1/2}P^{1/2}$.

6.2 Differentially Private Observer Design

Similarly to the situations considered in the previous chapters, suppose that we can observe a privacy-sensitive discrete-time signal $u := \{u_t\}_{t \in \mathbb{N}}$, with $u_t \in \mathsf{U} = \mathbb{R}^m$ for some positive integer m, and that we wish to process this observed signal to provide an estimate of another discrete-time signal denoted $x := \{x_t\}_{t \in \mathbb{N}}$, with $x_t \in \mathsf{X} = \mathbb{R}^n$ for some positive integer n, while providing a differential privacy guarantee for the input signal u. We assume in this chapter that u represents a collective data stream, in the sense of Sect. 2.3.1. The adjacency relations that we consider here to define the differential privacy guarantees for u, which we repeat for convenience, are the geometrically decaying deviation (2.6), i.e.,

$$\mathrm{Adj}(u, u') \text{ iff } \exists t_0, \text{ s.t. } \begin{cases} u_t = u'_t, t < t_0, \\ |u_t - u'_t|_p \leq K_p \, \alpha^{t-t_0}, t \geq t_0, \end{cases} \tag{6.1}$$

for $p = 1$ or 2 and some given constants $K_p > 0, 1 > \alpha \geq 0$, as well as the less restrictive relation (2.7), i.e.,

$$\text{Adj}(u, u') \text{ iff } \|u - u'\|_p \leq B_p, \tag{6.2}$$

for some constant $B_p > 0$.

Moreover, we assume publicly known a state-space model relating the signals u and x, as follows

$$x_{t+1} = f_t(x_t) + w_t, \tag{6.3}$$
$$u_t = g_t(x_t) + v_t, \tag{6.4}$$

where w_t, v_t are noise signals representing modeling and measurement errors, and f_t and g_t are known $\mathscr{C}^1$ functions, which, in contrast to the set-up of Chap. 5, can now be nonlinear. As usual, x_t could represent an aggregate state for a population of privacy-sensitive individuals, whose activities are observed via the signal u. For example, recalling the scenario of Sect. 3.5, x_t might be the density at period t of vehicles or pedestrians at a finite number of spatial locations, and u_t might be counts provided by motion detectors at a subset of these locations over time.

Our goal is to publish an estimate z_t of x_t, computed from u_t by a model-based observer, which, in the absence of privacy constraint, is assumed to take the following form (Sontag and Wang 1997)

$$z_{t+1} = f_t(z_t) + h_t(z_t, u_t - g_t(z_t)), \tag{6.5}$$

with, for each t in $\mathbb{N}$, $h_t : \mathsf{X} \times \mathsf{Y} \to \mathsf{X}$ a $\mathscr{C}^1$ function such that $h_t(x, 0) = 0$. In particular, if the measurement u_t at time t perfectly agrees with the prediction $g_t(z_t)$, then we predict the next step $f_t(z_t)$ based only on the model of the dynamics. We initialize (6.5) with some estimate z_0 of x_0. Note that (6.5) could describe an observer for a model (6.3)–(6.4) that has already been transformed under a suitable change of coordinates to a form that facilitates observer design, e.g., an observability canonical form (Gauthier and Kupka 2001; Isidori 2017). With straightforward modifications to our arguments, the "prediction" form (6.5) could also be replaced by an observer using the most recent observations

$$z_0 = \bar{z}_0 + h_0(\bar{z}_0, u_0 - g_t(\bar{z}_0)), \quad \text{for some estimate } \bar{z}_0 \text{ of } x_0,$$
$$z_{t+1} = f_t(z_t) + h_{t+1}(z_t, u_{t+1} - g_t(f_t(z_t))), \quad \text{for } t \geq 0. \tag{6.6}$$

The functions f_t, g_t and h_t are assumed to be publicly available, or at least could be potentially known to an adversary trying to make inferences about u based on z. To produce a differentially private version of z, we need to introduce noise in the design of the observer. The basic mechanisms of Chap. 2 introduce this privacy-preserving noise directly on the input signal u (input perturbation), or on the output signal z (output perturbation). In the first case, the observer can then attempts to smooth out the input noise, whereas in the second case, the sensitivity of the observer should be controlled. As in previous chapters, different designs offer different trade-offs between accuracy or convergence speed of the observer and privacy level offered, see for example the discussion in Sect. 5.4.4.

Remark 6.1 We do not provide here nor use any model of the noise signals w and v in (6.3), (6.4), which are simply introduced as a device to explain the discrepancy between any measured signal y and the signals that can be predicted by a deterministic model $x_{t+1} = f_t(x_t), u_t = g_t(x_t)$.

6.3 Input and Output Perturbation Mechanisms

Recall from Theorem 2.3 that we can obtain a differentially private signal at the output of a system G by adding noise with standard deviation proportional to $\Delta_1 G / \epsilon$ or to $\kappa_{\delta,\epsilon} \Delta_2 G$. For the input perturbation mechanism, see Sect. 2.3.2, the system G in Theorem 2.3 is taken to be the identity, with ℓ_1- and ℓ_2-sensitivity for the adjacency relation (6.1) equal to $K_1/(1-\alpha)$ and $K_2/\sqrt{1-\alpha^2}$ respectively, and B_1 and B_2 for (6.2). Note that for α close to 1, $1/\sqrt{1-\alpha^2}$ is significantly smaller than $1/(1-\alpha)$, so that if we are willing to accept some $\delta > 0$ in the privacy guarantee and to use the 2-norm in the adjacency relation (2.6), we can obtain much better accuracy by using the ℓ_2-sensitivity. Once the privacy-preserving noise is added at the input, the observer can then be designed according to any desired methodology and should try to mitigate the effect of this artificial noise in addition to the usual measurement error. For example, one can design an extended Kalman filter or any more computationally intensive nonlinear filter, such as a particle filter (Särkkä 2013), without necessarily restricting the design space to the form (6.5).

As noted in previous chapters, the input perturbation mechanism is attractive for its simplicity and the fact that sensitive data can be made differentially private at the source. It might also perform sufficiently well, especially with low privacy level requirements (high ϵ, δ) or low dimensional input signals. However, the noise added to u might be unnecessarily large, among other reasons because it is not tailored to the task of estimating the state x of the model (6.3)–(6.4), and does not take into account the temporal correlations between samples of the signal u captured by this model. Significant noise at the input of the observer can also lead to poor performance, i.e., slow convergence and large errors in the state estimate, or even perhaps divergence of the estimate from the true state trajectory, since the convergence of nonlinear observers is often local. In addition, characterizing the output error (state estimation error) due to the privacy-preserving noise requires understanding how this noise is transformed when passing through the nonlinear observer, which is significantly more difficult than for linear systems. For example, for nonlinear systems, the noise distribution at the output can become multimodal and non-zero mean, and hence the observer could produce a systematically biased estimate that could be hard to correct.

With output perturbation, a privacy-preserving noise signal proportional to the sensitivity of the observer G is added at its output. Computing the sensitivity of G, or in practice upper bounding it, should be done as accurately as possible to reduce the conservatism of the approach, a potentially difficult task for nonlinear systems. On the other hand the output noise does not impact any stability or bias analysis of the observer G. As discussed in previous chapters, for example in Sect. 5.4, we should then try to design an observer that has both good tracking performance for the state trajectory and low sensitivity, in order to minimize the level of privacy-preserving noise necessary at the output. These two desired properties are essentially in conflict. As in the previous chapters, see for example Sect. 5.4.3, a second stage filter can also be added to smooth out the privacy-preserving noise. Ideally, the resulting two-stage architecture should be optimized by following the methodology of Sect. 3.2. However, this is still a topic of research and hence not addressed in this chapter.

Example 6.1 Consider the memoryless system $u_t \mapsto \phi(u_t) := u_t^2$, which could be a simple state estimator for a measurement model $u_t = \sqrt{x_t}$ in (6.4), not taking the dynamics (6.3) into account. Consider the adjacency relation (6.1) for $\alpha = 0$, so that we have a deviation at some (unknown) single time period t_0 of at most K between adjacent signals u_t and $\tilde{u}_t$. For the input perturbation scheme and the Gaussian mechanism, assuming for simplicity that $\kappa_{\delta,\epsilon} = 1$, the signal $z_t = (u_t + K\xi_t)^2 = u_t^2 + 2Ku_t\xi_t + K^2\xi_t^2$ is differentially private when ξ is a standard Gaussian white noise. The privacy-preserving noise at the input induces a systematic bias at the output between z_t and u_t^2 equal to $\mathbb{E}[2Ku_t\xi_t + K^2\xi_t^2] = K^2$. Since K is assumed publicly known, in this case the bias can be compensated and a possibly better approximation of ϕ that is still differentially private is $z_t' = (u_t + K\xi_t)^2 - K^2$. One can verify that the variance of the remaining error is $e_t' = \mathbb{E}[(z_t' - u_t^2)^2] = 4K^2u_t^2 + 2K^4$.

Suppose we know in addition that $u_t \in [0, 1]$ for all $t \geq 0$. Then we can bound the sensitivity of the memoryless system as

$$\Delta_2\phi = |u_{t_0}^2 - \tilde{u}_{t_0}^2| = |u_{t_0} - \tilde{u}_{t_0}||u_{t_0} + \tilde{u}_{t_0}| \leq 2|u_{t_0} - \tilde{u}_{t_0}| \leq 2K. \quad (6.7)$$

Hence, the signal $z_t'' = u_t^2 + 2K\xi_t$ is also differentially private and unbiased, with ξ a standard white Gaussian noise as before. The variance of the error is $e_t'' = \mathbb{E}[(z_t'' - u_t^2)^2] = 4K^2$, which is smaller than the worst case value $4K^2 + 2K^4$ for e_t'. However, e_t'' is larger than e_t' as soon as $u_t < \sqrt{1 - K^2/2}$, the typical case since K should be much less than 1, otherwise both the input and output mechanisms essentially destroy the signal. The upper bound (6.7) on the sensitivity is conservative in order to be independent of the actual values of the sensitive signal u, which is necessary when noise proportional to the global sensitivity is used to provide a differential privacy guarantee.

In the rest of this chapter we focus on the output perturbation mechanism, i.e., adding privacy-preserving white noise directly on the signal z constructed from (6.5). In this case, the design of the function h_t for the observer must satisfy two requirements. First, we need to enforce appropriate asymptotic convergence of z toward x, which is the standard observer design problem, i.e., satisfy

$$|z_t - x_t|_\mathsf{X} \to 0 \text{ as } t \to \infty,$$

for some norm $|\cdot|_X$ on X, when the noise signals w and v in (6.3) and (6.4) are identically zero. Second, in order to appropriately set the magnitude of the noise added to the output signal z, we should consider the result of perturbing the input u of the observer (6.5) to an adjacent signal $\tilde{u}$ (for the adjacency relation (6.1) or (6.2)), leading to the new estimate

$$\tilde{z}_{t+1} = f_t(\tilde{z}_t) + h_t(\tilde{z}_t, \tilde{u}_t - g_t(\tilde{z}_t)).$$

We then need to control and bound explicitly the magnitude of the deviations between z and $\tilde{z}$, by bounding the signal norm $\|z - \tilde{z}\|_p$ for $p = 1$ or $p = 2$ in order to use the Laplace or Gaussian mechanism.

Here we handle both of these requirements by using contraction analysis to design a convergent observer and at the same time bound its sensitivity to variations in the measured signal u. A motivation for this approach is the exponential convergence of trajectories of contractive systems toward each other, which provides a degree of robustness against input disturbances (Lohmiller and Slotine 1998; Sontag and Willems 2010; Chung et al. 2008) and, as a consequence, sensitivity bounds for variations in input data streams u. The next section provides some background on contraction analysis, before we can describe our design approach in Sect. 6.5.

6.4 Contracting Systems

Contraction analysis is an "incremental" stability analysis methodology for dynamical systems emphasizing convergence of trajectories toward each other (Lewis 1949; Hartman 1982; Lohmiller and Slotine 1998). Contraction and incremental stability analysis have seen significant developments in the past two decades (Lohmiller and Slotine 1998; Angeli 2000; Aghannan and Rouchon 2003; Pavloc et al. 2006; Chung et al. 2008; Russo et al. 2010; Sontag and Willems 2010; Forni and Sepulchre 2014), and we refer the reader to the recent paper by Forni and Sepulchre (2014) for a comparison of different variations that have emerged and additional references. Although our focus here is on using contraction analysis, let us note that other techniques to establish incremental stability could be useful as well. Indeed, let us view our observer as a dynamical system $z = \mathscr{F}_{z_0}(u)$ linking the input signal u to the output signal z (and parametrized by the initial condition z_0) and suppose that we can establish an *incremental input-to-state stability* property of the form

$$|[\mathscr{F}_{z_0}(u)]_t - [\mathscr{F}_{\tilde{z}_0}(\tilde{u})]_t| \le \beta(|z_0 - \tilde{z}_0|, t) + [\mathscr{G}u]_t - [\mathscr{G}\tilde{u}]_t$$

for some function β of class $\mathscr{K}\mathscr{L}$, and some (simpler, scalar-valued) system $\mathscr{G}$ for which we can bound the ℓ_p-incremental variations as follows

$$\|\mathscr{G}u - \mathscr{G}\tilde{u}\|_p \le \gamma(\|u - \tilde{u}\|_p),$$

for $p = 1$ or 2 and some function γ of class $\mathcal{K}$. Then, taking first $u_t = \tilde{u}_t = g(x_t)$ for x, u signals following the model (6.3)–(6.4) with $w = v = 0$, and remarking that $x = \mathcal{F}_{x_0}(u)$ since the second term in (6.5) is then zero, we get that

$$|z_t - x_t| \leq \beta(|z_0 - x_0|, t),$$

which implies that the observer state z converges to x in the situation where the noiseless model would explain the signal u perfectly. Moreover if u, $\tilde{u}$ are adjacent signals according to (6.2) and the observer initialization z_0 is fixed independently of the input u, then we have

$$\|\mathcal{F}_{z_0}(u) - \mathcal{F}_{z_0}(\tilde{u})\|_p \leq \gamma(B_p),$$

which provides a bound on the ℓ_p-sensitivity that can be used to set the level of privacy-preserving Laplace or Gaussian noise. The challenge in practice consists in computing explicitly a function γ that is as tight as possible, in order to avoid adding an excessive amount of noise on the output or unnecessarily reducing the privacy level guarantee.

In this section we review some aspects of the contraction analysis methodology for discrete-time systems and prove some results that we rely on to design differentially private observers with output perturbation. Our discussion is self-contained and in particular we provide explicit bounds on distances between trajectories, which are necessary to precisely set the level of privacy-preserving noise for output perturbation.

6.4.1 Basic Results

Our first task is to define formally what it means for a system to be contractive and to present tools that can be used to establish quantitative properties for the trajectories of such systems. Consider a discrete-time system

$$x_{t+1} = f_t(x_t), \tag{6.8}$$

with $f_t : \mathsf{X} \to \mathsf{X}$ a $\mathscr{C}^1$ function, for all $t \in \mathbb{N}$. Let us denote by $\phi(t; t_0, x_0)$ the value at time $t \geq t_0$ of the solution of (6.8) taking the value x_0 at time t_0. A forward invariant set for the system (6.8) is a set $C \subset \mathsf{X}$ such that if $x_0 \in C$, then for all t_0 and all $t \geq t_0$, $\phi(t; t_0, x_0) \in C$. Although we assume here that $\mathsf{X} = \mathbb{R}^n$, it is useful to introduce here some language from differential geometry and view X more generally as an n-dimensional differentiable manifold (do Carmo 1992; Forni and Sepulchre 2014). For each point $x \in \mathsf{X}$, the tangent space to X at x, i.e., informally, the n-dimensional vector space of all tangent vectors to curves on X passing through x, is denoted $\mathsf{T}_x\mathsf{X}$. The tangent bundle of X is denoted $\mathsf{TX} := \cup_{x \in \mathsf{X}}\{x\} \times \mathsf{T}_x\mathsf{X}$, and is equipped with a time-varying family of norms $|\cdot|_{[x,t]}$, smoothly varying with x for each t, so that $|\cdot|_{[x,t]}$ is a norm on $\mathsf{T}_x\mathsf{X}$, for all $t \in \mathbb{N}$. For each $x, \tilde{x} \in \mathsf{X}$, let $\Gamma(x, \tilde{x})$ be the set of piecewise $\mathscr{C}^1$ curves joining x and $\tilde{x}$, i.e., functions $\gamma : [0, 1] \to \mathsf{X}$ with $\gamma(0) = x$, $\gamma(1) = \tilde{x}$. We define the (time-varying) length of such a curve γ by (Shen 2001; Forni and Sepulchre 2014)

$$L_t(\gamma) = \int_0^1 |\gamma'(r)|_{[\gamma(r),t]}dr,$$

where $\gamma'(r) := \frac{d\gamma}{dr}(r) \in \mathsf{T}_{\gamma(r)}\mathsf{X}$ is the tangent vector to the curve at the point $\gamma(r)$. We then have a notion of (time-varying) geodesic distance on X, defined as

$$d_t(x, \tilde{x}) = \inf_{\gamma \in \Gamma(x,\tilde{x})} L_t(\gamma), \quad \forall x, \tilde{x} \in \mathsf{X}. \tag{6.9}$$

Moreover, if the norms $|\cdot|_{[x,t]}$ are in fact independent of x, thus denoted $|\cdot|_{[t]}$, and if X is $\mathbb{R}^n$ or a convex subset of $\mathbb{R}^n$, then the infimum in (6.9) is achieved by straight lines $\gamma(r) = x + r(\tilde{x} - x)$ and $d_t(x, \tilde{x}) = |\tilde{x} - x|_{[t]}$ in (6.9). Finally, each function f_t in (6.8) is associated to a Jacobian $F_t(x) := \frac{\partial f_t}{\partial x}(x)$, which defines a linear map from $\mathsf{T}_x\mathsf{X}$ at time t to $\mathsf{T}_{f_t(x)}\mathsf{X}$ at time $t + 1$. As a result, for all vectors $v \in \mathsf{T}_x\mathsf{X}$,

$$|F_t(x)\,v|_{[f_t(x),t+1]} \le \|F_t(x)\|^{[x,t]}_{[f_t(x),t+1]}\,|v|_{[x,t]}, \tag{6.10}$$

where $\|\cdot\|^{[x,t]}_{[f_t(x),t+1]}$ denotes the norm induced by $|\cdot|_{[x,t]}$ and $|\cdot|_{[f_t(x),t+1]}$.

Definition 6.1 Let ρ be a nonnegative constant. The system (6.8) is said to be ρ-*contracting* for the norms $|\cdot|_{[x,t]}$ on a forward invariant set $C \subset \mathsf{X}$ if for any $t_0 \in \mathbb{N}$ and any two initial conditions $x_0, \tilde{x}_0 \in C$, we have, for all $t \ge t_0$,

$$d_t(\phi(t; t_0, x_0), \phi(t; t_0, \tilde{x}_0)) \le \rho^{t-t_0} d_{t_0}(x_0, \tilde{x}_0). \tag{6.11}$$

Let $\gamma_t \in \Gamma(x, \tilde{x})$ be a curve joining two points x and $\tilde{x}$ in X at a fixed time period t. Let $\gamma_t'(r)$ be the tangent vector to γ_t at the point $\gamma_t(r)$, for $r \in [0, 1]$. The curve γ_t is transported at time t by (6.8) to a curve γ_{t+1} joining $f_t(x)$ and $f_t(\tilde{x})$. Taking the derivative with respect to r in the equation $\gamma_{t+1}(r) = f_t(\gamma_t(r))$, we obtain an important *linear* relation between tangent vectors

$$\gamma_{t+1}'(r) = F_t(\gamma_t(r))\,\gamma_t'(r), \quad \forall r \in [0, 1], \forall t \ge 0. \tag{6.12}$$

The following fundamental theorem of contraction analysis is then a consequence of (6.12).

Theorem 6.1 *Let* $F_t = \frac{\partial f_t}{\partial x}$ *be the Jacobian of* f_t, *for all* $t \ge 0$. *A sufficient condition for the system* (6.8) *to be* ρ-*contracting for the norms* $|\cdot|_{[x,t]}$ *on a forward invariant set* $C \subset \mathsf{X}$ *is that*

$$\|F_t(x)\|^{[x,t]}_{[f_t(x),t+1]} \le \rho, \quad \forall x \in C, \forall t \in \mathbb{N}. \tag{6.13}$$

Proof Consider a curve $\gamma_{t_0} : [0, 1] \to$ **X** in $\Gamma(x_0, \tilde{x}_0)$. This curve is transported by (6.8) to a sequence of curves $\gamma_{t_0+1}, \gamma_{t_0+2}, \ldots$, i.e., $\gamma_{t+1}(r) = f_t(\gamma_t(r))$, for all $r \in [0, 1]$, with γ_t joining $\phi(t; t_0, x_0)$ and $\phi(t; t_0, \tilde{x}_0)$. We have, for all $t \geq t_0$, using (6.12)

$$L_{t+1}(\gamma_{t+1}) = \int_0^1 |\gamma'_{t+1}(r)|_{[\gamma_{t+1}(r), t+1]} dr = \int_0^1 |F_t(\gamma_t(r)) \gamma'_t(r)|_{[\gamma_{t+1}(r), t+1]} dr.$$

Now, using (6.10) and then the assumption (6.13)

$$L_{t+1}(\gamma_{t+1}) \leq \int_0^1 \|F_t(\gamma_t(r))\|_{[\gamma_{t+1}(r), t+1]}^{[\gamma_t(r), t]} |\gamma'_t(r)|_{[\gamma_t(r), t]} dr$$

$$\leq \rho \int_0^1 |\gamma'_t(r)|_{[\gamma_t(r), t]} dr = \rho L_t(\gamma_t), \tag{6.14}$$

and hence by immediate recursion, $L_t(\gamma_t) \leq \rho^{t-t_0} L_{t_0}(\gamma_{t_0})$. To conclude, let $\epsilon > 0$ and take the curve γ_{t_0} above to satisfy

$$L_{t_0}(\gamma_{t_0}) \leq (1 + \epsilon) d_{t_0}(x_0, \tilde{x}_0).$$

Then, since $\gamma_t \in \Gamma(\phi(t; t_0, x_0), \phi(t; t_0, \tilde{x}_0))$, we have

$$d_t(\phi(t; t_0, x_0), \phi(t; t_0, \tilde{x}_0)) \leq L_t(\gamma_t) \leq \rho^{t-t_0} L_{t_0}(\gamma_{t_0}) \leq (1 + \epsilon) \rho^{t-t_0} d_{t_0}(x_0, \tilde{x}_0). \tag{6.15}$$

Since this inequality is true for all $\epsilon > 0$, (6.11) holds. $\square$

As a result of Theorem 6.1 and the remarks on geodesic distances preceding Definition 6.1, we have the following contraction result when the case of state-independent norms.

Corollary 6.1 *With the notation defined as in Theorem 6.1, suppose that C is a convex forward invariant subset of $\mathbb{R}^n$ and that the norms $|\cdot|_{[x,t]}$ on the tangent spaces are independent of x and denoted $|\cdot|_t$. Let $\|\cdot\|_{t+1}^t$ be the matrix norm induced by $|\cdot|_t$ and $|\cdot|_{t+1}$. Then, if $\|F_t(x)\|_{t+1}^t \leq \rho$ for all $x \in C$ and for all $t \in \mathbb{N}$, we have*

$$|\phi(t; t_0, x_0) - \phi(t; t_0, \tilde{x}_0)|_t \leq \rho^{t-t_0} |x_0 - \tilde{x}_0|_{t_0}, \quad \forall x_0, \tilde{x}_0 \in C, \forall t \geq t_0.$$

We also specialize the result of Theorem 6.1 to situations where we can carry out explicit computations for the 1- and 2-norm, which are of particular interest for the Laplace and Gaussian mechanism.

Corollary 6.2 *With the notation defined as in Theorem 6.1, suppose that the norms on the tangent spaces are defined for all x and t by $|v|_{[x,t]} = |P_{[x,t]} v|_1$, where $P_{[x,t]} = \text{diag}(p_{[x,t]})$, with $p_{[x,t]}$ a vector with positive components $p_{[x,t],i}$. Hence, $|v|_{[x,t]} = \sum_{i=1}^n p_{[x,t],i} |v_i|$. Then the system is ρ-contracting for the associated distances on **X** if the following linear programs are feasible, for all $x \in C$ and $t \in \mathbb{N}$*

$$\sum_{i=1}^{n} p_{[f_t(x),t+1],i}|F_{t,ij}(x)| \leq \rho\, p_{[x,t],j}, \quad \forall 1 \leq j \leq n, \tag{6.16}$$

$$p_{[x,t],i},\, p_{[f_t(x),t+1],i} > 0, \quad \forall 1 \leq i \leq n. \tag{6.17}$$

In particular, if C is convex and if there exist positive vectors $p_{[t]}$ independent of x satisfying the above inequalities (6.16), (6.17) *for all x, t, then, with $P_{[t]} := diag(p_{[t]})$, $x_t := \phi(t; t_0, x_0)$, $\tilde{x}_t := \phi(t; t_0, \tilde{x}_0)$, we have*

$$|P_{[t]}(x_t - \tilde{x}_t)|_1 \leq \rho^{t-t_0}|P_{[t_0]}(x_0 - \tilde{x}_0)|_1, \quad \forall x_0, \tilde{x}_0 \in C, \forall t \geq t_0. \tag{6.18}$$

Proof The inequalities (6.16), (6.17) come from satisfying (6.13) for the 1-norms weighted by $R := P_{[x,t]}$ and $S := P_{[f_t(x),t+1]}$. The condition (6.13) is equivalent to the induced 1-norm of the matrix $S F_t(x) R^{-1}$ being less than ρ, and this matrix has entries $p_{[f_t(x),t+1],i} F_{t,ij}(x)/p_{[x,t],j}$. The induced 1-norm of an $n \times m$ matrix $A = [a_{ij}]_{i,j}$ is $\max_{1 \leq j \leq m} \sum_{i=1}^{n} |a_{ij}|$. The result follows from these facts.

Corollary 6.3 *With the notation defined as in Theorem 6.1, suppose that the norms on the tangent spaces are defined by $|v|_{[x,t]} = (v^T P_{[x,t]}v)^{1/2} = |P_{[x,t]}^{1/2}v|_2$, where $P_{[x,t]} \succ 0$, for all x and t. Then the system is ρ-contracting for the associated distances on X if the following LMIs are satisfied*

$$F_t(x)^T P_{[f_t(x),t+1]} F_t(x) \preceq \rho^2 P_{[x,t]}, \quad \forall x \in C, \forall t \in \mathbb{N}. \tag{6.19}$$

Suppose C is convex. If there exist matrices $P_{[t]} \succ 0$, $t \in \mathbb{N}$, independent of x, satisfying these LMIs, then we have

$$|P_{[t]}^{1/2}(x_t - \tilde{x}_t)|_2 \leq \rho^{t-t_0}|P_{[t_0]}^{1/2}(x_0 - \tilde{x}_0)|_2, \quad \forall x_0, \tilde{x}_0 \in C, \forall t \geq t_0, \tag{6.20}$$

where $x_t := \phi(t; t_0, x_0)$, $\tilde{x}_t := \phi(t; t_0, \tilde{x}_0)$. If there exist matrices $P_{[x,t]}$ satisfying (6.19) *and if there there exist 2 matrices $P_{\min} \succ 0$ with minimum eigenvalue $\lambda_{\min} > 0$ and $P_{\max} \succ 0$ with maximum eigenvalue $\lambda_{\max} > 0$ such that we have $\lambda_{min}I \preceq P_{\min} \preceq P_{[x,t]} \preceq P_{\max} \preceq \lambda_{max}I$, for all x, t, then*

$$|P_{\min}^{1/2}(x_t - \tilde{x}_t)|_2 \leq \rho^{t-t_0}|P_{\max}^{1/2}(x_0 - \tilde{x}_0)|_2, \quad \forall x_0, \tilde{x}_0 \in C, \forall t \geq t_0,$$

and hence

$$|x_t - \tilde{x}_t|_2 \leq \rho^{t-t_0}\sqrt{\frac{\lambda_{\max}}{\lambda_{\min}}}|x_0 - \tilde{x}_0|_2.$$

Proof This is a corollary of Theorem 6.1, since satisfying (6.13) for the norm induced by the weighted 2-norms with matrices $P_{[x,t]}$ and $P_{[f_t(x),t+1]}$ can be written $v^T F_t(x)^T P_{[f_t(x),t+1]} F_t(x)v \leq \rho^2 v^T P_{[x,t]}v$, for all v in $\mathbb{R}^n$. The second part uses the following fact

$$\int_0^1 \sqrt{\gamma'(r) P_{\max} \gamma'(r)} dr \geq L_t(\gamma) \geq \int_0^1 \sqrt{\gamma'(r) P_{\min} \gamma'(r)} dr,$$

since $L_t(\gamma) = \int_0^1 \sqrt{\gamma'(r) P_{[\gamma(r),k]} \gamma'(r)} dr$. Moreover

$$\int_0^1 \sqrt{\gamma'(r) P_{\min} \gamma'(r)} dr \geq |P_{min}^{1/2}(x - \tilde{x})|_2 \text{ if } \gamma \in \Gamma(x, \tilde{x}),$$

since for a constant norm on $\mathbb{R}^n$ the geodesic curves are straight lines. Finally, referring to the argument leading to (6.15), we get

$$|P_{min}^{1/2}(x_t - \tilde{x}_t)|_2 \leq L_t(\gamma_t) \leq \rho^{t-t_0} L_{t_0}(\gamma_{t_0}) \leq \rho^{t-t_0} |P_{\max}^{1/2}(x_0 - \tilde{x}_0)|_2.$$

Remark 6.2 The first part of Corollary 6.3 is the classical contraction result (Lohmiller and Slotine 1998), in discrete time, for norms associated with an inner product (Riemannian structure on X). Using state-dependent matrices $P_{[x,t]}$ enlarges the set of systems for which we can prove contraction, but here we also need to explicitly bound the Euclidean distances $|x_t - \tilde{x}_t|_2$, not just general geodesic distances, in order to evaluate the level of noise necessary for the Gaussian mechanism.

6.4.2 Effect of Disturbances on Contractive Systems

To compute ℓ^1 and ℓ^2-sensitivities, we need to bound the size of trajectory deviations of contracting systems subject to disturbances. Qualitatively, the exponential convergence of trajectories of a contracting system provides some robustness against disturbances (Khalil 2002; Lohmiller and Slotine 1998; Sontag and Willems 2010). However, to precisely set the level of privacy-preserving noise, quantitative worst case bounds on the ℓ^1 or ℓ^2-norms of the trajectory deviations are needed. Hence, consider a system

$$x_{t+1} = f_t(x_t, \pi_t(x_t)), \tag{6.21}$$

where $\pi_t : X \to P := \mathbb{R}^p$, for some integer p, represents a $\mathscr{C}^1$ disturbance signal, and for all $t \geq 0$, $f_t : X \times P \to X$ is $\mathscr{C}^1$. We equip the tangent spaces of the product manifold $X \times P$ with time-varying norms assumed for simplicity to be fixed for the disturbance part, i.e., $|(v, w)|_{[(x,\pi),t]} = |v|_{[x,t]} + |w|_P$, for a fixed norm $| \cdot |_P$. The nominal system under zero disturbance is

$$\bar{x}_{t+1} = f_t(\bar{x}_t, 0). \tag{6.22}$$

We denote $\frac{\partial f_t}{\partial x}$ and $\frac{\partial f_t}{\partial \pi}$ the Jacobian matrices of $f_t(x, \pi)$ with respect to the components of x and π respectively. For $r \in [0, 1]$, denote by $\phi(t; r, t_0, x_0)$ the iterates

$$x_{t+1} = f_t(x_t, r\,\pi_t(x_t)), \tag{6.23}$$

starting from x_0 at time t_0. Note that (6.21) corresponds to $r = 1$ and (6.22) to $r = 0$.
Let us also define

$$J_t(x; r) := \frac{\partial f_t}{\partial x}(x, r\,\pi_t(x)) + r\frac{\partial f_t}{\partial \pi}(x, r\,\pi_t(x))\frac{\partial \pi_t}{\partial x}(x), \ \forall x \in \mathsf{X}, \forall r \in [0, 1]. \tag{6.24}$$

For all x in X, denote $x_+^{t,r} := f_t(x, r\,\pi_t(x))$. Formally, the "differential" maps (6.24) are from $\mathsf{T}_{[x,t]}\mathsf{X}$ to $\mathsf{T}_{[x_+^{t,r},t+1]}\mathsf{X}$, with the corresponding induced norms $\| \cdot \|_{[x_+^{t,r},t+1]}^{[x,t]}$. We then have the following result.

Theorem 6.2 *Consider a trajectory $\bar{x}_t := \phi(t; 0, t_0, \bar{x}_0)$ for (6.22) starting from $\bar{x}_0$ and a trajectory $x_t := \phi(t; 1, t_0, x_0)$ for the perturbed system (6.21) starting from x_0. Suppose that there exists a sequence $\{M_t\}_{t \geq t_0}$ such that*

$$\left| \frac{\partial f_t}{\partial \pi}(x, r\,\pi_t(x))\,\pi_t(x) \right|_{[x_+^{t,r},t+1]} \leq M_t, \ \ \forall r \in [0, 1], \forall x \in C, \forall t \geq t_0, \tag{6.25}$$

and that

$$\|J_t(x; r)\|_{[x_+^{t,r},t+1]}^{[x,t]} \leq \rho, \ \ \ \forall r \in [0, 1], \forall x \in C, \forall t \geq t_0, \tag{6.26}$$

where C is a forward invariant set for (6.23), for all $r \in [0, 1]$. Then we have, for all $t \geq t_0$, and the distances d_t defined in (6.9),

$$d_t(\bar{x}_t, x_t) \leq \rho^{t-t_0}d_{t_0}(\bar{x}_{t_0}, x_{t_0}) + \sum_{l=0}^{t-t_0-1} \rho^l M_{t-1-l}.$$

Remark 6.3 In the case of additive disturbances on $\mathsf{X} = \mathsf{P} = \mathbb{R}^n$, i.e.,

$$f_t(x, \pi_t(x)) = \tilde{f}_t(x) + \pi_t(x), \tag{6.27}$$

with a fixed norm $| \cdot |$ on $\mathbb{R}^n$, the condition (6.25) can be written more simply $\sup_{x \in C} |\pi_t(x)| \leq M_t$, i.e., M_t is a bound on the disturbance term.

Remark 6.4 Note that if the disturbance π_t does not depend on x, then (6.24) reads $J_t(x; r) := \frac{\partial f_t}{\partial x}(x, r\,\pi_t)$ and (6.26) is a type of contraction condition on the perturbed system. If moreover the perturbation is in fact additive as in (6.27), then (6.26) simply asks that the Jacobian of the nominal system $\tilde{f}_t$ satisfy the contraction assumption.

Proof Consider a curve $\gamma_{t_0} \in \Gamma(\bar{x}_0, x_0)$, i.e., such that $\gamma_{t_0}(0) = \bar{x}_0$ and $\gamma_{t_0}(1) = x_0$, transported by (6.23) to the sequence

$$\gamma_t(r) = \phi(t; r, t_0, \gamma_{t_0}(r)), \ \forall r \in [0, 1], \forall t \geq t_0.$$

Then, for $t \geq t_0$, we have $\gamma_t \in \Gamma(\bar{x}_t, x_t)$, where $\bar{x}_t := \phi(t; 0, t_0, \bar{x}_0)$ and $x_t := \phi(t; 1, t_0, x_0)$. Following the idea of the proof of Theorem 6.1, define $\gamma_t'(r) := \frac{d}{dr}\phi(t; r, t_0, \gamma_{t_0}(r))$, so that we have, for all t and all $r \in [0, 1]$

$$\gamma_{t+1}'(r) = J_t(\gamma_t(r); r)\, \gamma_t'(r) + \frac{\partial f_t}{\partial \pi}(\gamma_t(r), r\,\pi_t(\gamma_t(r)))\, \pi_t(\gamma_t(r)),$$

which implies, by (6.26) and (6.25),

$$|\gamma_{t+1}'(r)|_{[\gamma_{t+1}(r), t+1]} \leq \rho\, |\gamma_t'(r)|_{[\gamma_t(r), t]} + M_t, \quad \forall r \in [0, 1], \forall t \geq t_0,$$

and by integration over $r \in [0, 1]$

$$L_{t+1}(\gamma_{t+1}) \leq \rho L_t(\gamma_t) + M_t, \quad \forall t \geq t_0.$$

By the comparison lemma (Lakshmikantham and Trigiante 2002), we then have that $L_t(\gamma_t) \leq u_t$ for u_t satisfying the linear scalar dynamics

$$u_{t_0} = L_{t_0}(\gamma_{t_0}), \quad u_{t+1} = \rho\, u_t + M_t, \quad \forall t \geq t_0.$$

Hence, $L_t(\gamma_t) \leq \rho^{t-t_0} u_{t_0} + \sum_{l=0}^{t-t_0-1} \rho^l M_{t-1-l}$. As in the end of the proof of Theorem 6.1, we can then choose γ_{t_0} so that $L_{t_0}(\gamma_{t_0})$ is arbitrarily close to $d_{t_0}(\bar{x}_0, x_0)$, and then use $d_t(\bar{x}_t, x_t) \leq L_t(\gamma_t)$ to conclude.

We can now make convergence assumptions on the bounding sequence $\{M_t\}_{t \geq t_0}$ in (6.25) to state more concrete results. The following corollaries follow by standard calculations (Khalil 2002) on the sequence u_t introduced at the end of the proof of Theorem 6.2.

Corollary 6.4 *Let $1 \leq p \leq \infty$ be an integer. Suppose that $\{M_t\}_{t \geq t_0}$ in (6.25) is a sequence in ℓ^p, with norm $\|M\|_p$. Then, with the notation and assumptions of Theorem 6.2, if $\rho < 1$, there exists a class $\mathscr{K}$ function $\beta : \mathbb{R}_+ \to \mathbb{R}_+$ such that*

$$\left(\sum_{t=t_0}^{\infty} d_t(\bar{x}_t, x_t)^p\right)^{1/p} \leq \beta(d_{t_0}(x_0, \bar{x}_0)) + \frac{\|M\|_p}{1 - \rho}, \tag{6.28}$$

where, for $p = \infty$, the left-hand side of the inequality is interpreted as usual as $\sup_{t \geq t_0} d_t(\bar{x}_t, x_t)$.

By further restricting the class of disturbances, we get slightly tighter bounds on the deviations for $p \geq 2$.

Corollary 6.5 *Let $1 \leq p \leq \infty$ be an integer. Suppose that $\{M_t\}_{t \geq 0}$ in (6.25) satisfies the following condition:*

$$\exists K \geq 0,\, 1 > \alpha \geq 0,\, \text{and } t_0 \in \mathbb{N} \text{ s.t. } M_t = \begin{cases} 0, & \text{if } t < t_0, \\ K\alpha^{t-t_0}, & \text{if } t \geq t_0. \end{cases} \tag{6.29}$$

Then, with the notation and assumptions of Theorem 6.2, for $t \geq t_0$,

$$d_t(\bar{x}_t, x_t) \leq \rho^{t-t_0} d_{t_0}(\bar{x}_0, x_0) + K \frac{\rho^{t-t_0} - \alpha^{t-t_0}}{\rho - \alpha}.$$

Hence, if $\rho < 1$,

$$\sum_{t=t_0}^{\infty} d_t(\bar{x}_t, x_t) \leq \frac{1}{1-\rho} d_{t_0}(\bar{x}_0, x_0) + \frac{K}{(1-\rho)(1-\alpha)},$$

and for any $p \geq 2$, there exists a class $\mathcal{K}$ function $\beta : \mathbb{R}_+ \to \mathbb{R}_+$ such that

$$\left(\sum_{t=t_0}^{\infty} d_t(\bar{x}_t, x_t)^p \right)^{1/p} \leq \beta(d_{t_0}(\bar{x}_0, x_0)) + \frac{K}{|\rho - \alpha|} \left(\sum_{t=0}^{\infty} |\rho^t - \alpha^t|^p \right)^{1/p}. \quad (6.30)$$

Remark 6.5 If the norms on TX are given by weighted 1 and 2-norms as in Corollaries 6.2 and 6.3, then condition (6.26) corresponds to the feasibility of a family of linear programs or LMIs, and if moreover C is convex and the weight matrices in these norms are independent of x, then we can replace the distances $d_t(\bar{x}_t, x_t)$ in (6.28), (6.30) by $|P_{[t]}(\bar{x}_t - x_t)|_1$ or $|P_{[t]}^{1/2}(\bar{x}_t - x_t)|_2$ as in (6.18), (6.20).

6.5 Differentially Private Observers with Output Perturbation

Let us now return to our initial differentially private observer design problem with output perturbation. Two adjacent measured signals y and $\tilde{y}$ produce distinct observer state trajectories z and $\tilde{z}$ by (6.5), i.e., such that

$$z_{t+1} = f_t(z_t) + h_t(z_t, y_t - g_t(z_t)), \quad (6.31)$$
$$\tilde{z}_{t+1} = f_t(\tilde{z}_t) + h_t(\tilde{z}_t, y_t - g_t(\tilde{z}_t) + \pi_t), \quad (6.32)$$

where $\pi_t = \tilde{y}_t - y_t$. Recalling the discussion at the beginning of Sect. 6.4, we can now attempt to choose the functions h_t to design a contractive observer, while at the same time minimizing the "gain" of the map $\pi \to z$. First, contraction provides a notion of convergence for the observer. Namely, if the model (6.3), (6.4) were valid under no modeling noise assumptions (zero v, w), then any the sequence x satisfying (6.3), (6.4) would also satisfy the dynamics (6.31) (since $y_t = g(x_t)$), and the trajectories x, z would converge exponentially toward each other. In particular, any initial difference between z_0 (state estimate used to initialize the observer) and x_0 (true initial state) would eventually be forgotten. Second, the results of Sect. 6.4.2 give us tools to bound the sensitivity of contractive observers, i.e., the deviations between z and $\tilde{z}$ above, and hence a means to set the level of privacy-preserving noise for the Laplace or Gaussian output perturbation mechanism.

Given two measured signals y and $\tilde{y}$, define the notation $v_t^{y,\tilde{y}}(x;r) := y_t - g_t(x) + r\pi_t = (1-r)y_t + r\tilde{y}_t - g_t(x)$ and

$$J_t^{y,\tilde{y}}(x;r) = \frac{\partial f_t}{\partial x}(x) + \frac{\partial h_t}{\partial x}(x, v_t^{y,\tilde{y}}(x;r)) - \frac{\partial h_t}{\partial y}(x, v_t^{y,\tilde{y}}(x;r))\frac{\partial g_t}{\partial x}(x). \quad (6.33)$$

The proof of the following proposition follows immediately from Theorem 6.2 and Remark 6.4.

Proposition 6.1 *Consider the observer* (6.5), *and two measured signals* y, $\tilde{y}$ *producing respectively the trajectories* z, $\tilde{z}$, *assuming the same initial condition* $z_0 = \tilde{z}_0$ *to initialize the observer. Suppose that we have the bound*

$$\|J_t^{y,\tilde{y}}(x;r)\|_{[x_+^{t,r},t+1]}^{[x,t]} \leq \rho, \quad \forall r \in [0,1], \forall x \in C, \forall t \in \mathbb{N}, \quad (6.34)$$

where $J_t^{y,\tilde{y}}$ *is defined by* (6.33), $x_+^{t,r} := f_t(x) + h_t(x, v_t^{y,\tilde{y}}(x;r))$ *and* C *is a set containing* z_0, *which is forward invariant for the observer* (6.31) *for any input signal* $(1-r)y + r\tilde{y}$, $r \in [0,1]$. *Suppose moreover that*

$$\sup_{x \in C, r \in [0,1]} \left| \frac{\partial h_t}{\partial y}(x, v_t^{y,\tilde{y}}(x;r))(\tilde{y}_t - y_t) \right|_{[x_+^{t,r},t+1]} \leq M_t, \quad \forall t \in \mathbb{N}. \quad (6.35)$$

Then, we have, for the geodesic distances d_t *associated to the norms* $|\cdot|_{[x,t]}$

$$d_t(z_t, \tilde{z}_t) \leq \sum_{l=0}^{t-1} \rho^l M_{t-1-l}.$$

In the rest of this section, we illustrate how the general result of Proposition 6.1 can be applied in some relatively simple situations were calculations can be carried out efficiently to obtain the final privacy-preserving observer. We focus in particular on problems where a Luenberger-type observer can be used to estimate the state

$$z_{t+1} = f_t(z_t) + H_t \times (y_t - g_t(z_t)), \quad (6.36)$$

where H_t represents a $n \times m$ matrix to design. In other words, we set $h_t(x, y) = H_t\, y$, a linear relation that does not depend on the value of x. Then the right-hand side of the expression (6.33) reads simply $\frac{\partial f_t}{\partial x}(x) - H_t \frac{\partial g_t}{\partial x}(x)$ and becomes in particular independent of r and y, $\tilde{y}$. Next, fix a norm $|\cdot|_{\mathsf{X}}$ on TX, independent of x, t, and a p-norm $|\cdot|_p$ on Y, and let

$$\bar{H}_{\mathsf{X}}^p := \sup_t \|H_t\|_{\mathsf{X}}^{\mathsf{Y}}.$$

Then we can take $M_t = \bar{H}_{\mathsf{X}}^p |y_t - \tilde{y}_t|_p$ in (6.35). This leads to the following corollary of Proposition 6.1, similar to the Corollaries 6.4 and 6.5, which we will use next in the illustrative examples. We introduce the notation $\|v\|_{p,\mathsf{X}} := \left(\sum_{k=0}^{\infty} |v_t|_{\mathsf{X}}^p\right)^{1/p}$, for $1 \leq p \leq \infty$.

Corollary 6.6 *Consider the observer* (6.36), *and two measured signals* $y, \tilde{y}$ *producing respectively the trajectories* $z, \tilde{z}$, *assuming the same initial condition* $z_0 = \tilde{z}_0$ *to initialize the observer. Fix the norms* $|\cdot|_{\mathsf{X}}$, *on* TX, *independent of* x, t. *Suppose that we have the bound*

$$\left\| \frac{\partial f_t}{\partial x}(x) - H_t \frac{\partial g_t}{\partial x}(x) \right\|_{\mathsf{X}} \leq \rho, \quad \forall x \in C, t \in \mathbb{N}, \tag{6.37}$$

for some constant $\rho < 1$, *where* C *is a set containing* z_0 *and forward invariant for* (6.31) *for any input signal* $y + (1-r)\tilde{y}$, $r \in [0, 1]$. *Then, if the signals* $y, \tilde{y}$ *are adjacent according to* (6.2), *we have, for the same value of* p,

$$\|z - \tilde{z}\|_{p,\mathsf{X}} \leq \frac{B_p \, \bar{H}_{\mathsf{X}}^p}{1 - \rho}. \tag{6.38}$$

Moreover, if the signals $y, \tilde{y}$ *are in fact adjacent according to* (6.1), *we have more precisely, for the same value of* p,

$$\|z - \tilde{z}\|_{p,\mathsf{X}} \leq \frac{K_p \, \bar{H}_{\mathsf{X}}^p}{|\rho - \alpha|} \left(\sum_{t=0}^{\infty} |\rho^t - \alpha^t|^p\right)^{1/p}. \tag{6.39}$$

Remark 6.6 For the adjacency relation (6.1) with $p = 1$, both (6.39) and (6.38) give the same upper bound $\frac{K_p \, \bar{H}_{\mathsf{X}}^p}{(1-\rho)(1-\alpha)}$, since B_p in (6.2) and (6.38) represents in fact $K_p/(1 - \alpha)$ when the adjacency relation is (6.1). But (6.39) is useful for the tighter adjacency relation (6.1) and $p = 2$, in order to design a mechanisms with a lower level of Gaussian noise compared to using the adjacency relation (6.2).

In Corollary 6.6, the choice of the gain matrices H_t has an impact both on ρ and on the ℓ_p-sensitivity bound. Increasing the gains H_t can help decrease the contraction rate ρ to obtain a more rapidly converging observer, but at the same time it increases the sensitivity and thus the level of noise necessary for differential privacy. Hence, in general, we should try to achieve a reasonable contraction rate ρ with the smallest gain possible. We conclude this section with two more corollaries, describing concrete differentially private observers with output perturbation.

Corollary 6.7 *Let* $P = \mathrm{diag}(p)$, *with* $p_i > 0, 1 \leq i \leq n$, *and assume that the conditions of Corollary 6.6 are satisfied for the weighted* 1-*norm* $|Pv|_1 = \sum_{i=1}^{n} p_i |v_i|$ *on* X. *Consider the signal* $\hat{x}_t = z_t + \xi_t$, *where* z_t *is computed from* (6.36), *and* $\xi_{t,i}$ *are iid Laplace random variables with parameters* b/p_i, *for* $1 \leq i \leq n$, *where*

$$b = \frac{B_1 \sup_t \| P H_t \|_1}{\epsilon (1 - \rho)}. \tag{6.40}$$

Then this signal $\hat{x}_t$ is ϵ-differentially private for the adjacency relation (6.2) with $p = 1$, and for (6.1) with $p = 1$ when $B_1 = K_1/(1 - \alpha)$.

Proof From the bound (6.38) for $p = 1$, since $\| z - \tilde{z} \|_{1,\mathsf{X}} = \sum_{t=0}^{\infty} | P(z_t - \tilde{z}_t) |_1$ we deduce by Theorem 2.3 that $P z_t + \zeta_t$ is a differentially private signal, where ζ_k has Laplace distributed iid components with the parameter b. Hence $P^{-1}(P z_t + \zeta_t)$ is also differentially private by resilience to post-processing, see Sect. 1.3.3. We define $\xi_t = P^{-1} \zeta_t$ in the Corollary.

Corollary 6.8 *Let P be a positive definite matrix, and assume that the conditions of Corollary 6.6 are satisfied for the weighted 2-norm $|P^{1/2} v|_2$ on X. Consider the signal $\hat{x}_t = z_t + \xi_t$, where z_t is computed from (6.36), and ξ_t is a Gaussian white noise with covariance matrix $\sigma^2 P^{-1}$, where $\sigma = \kappa_{\delta,\epsilon} C_2 \sup_t \| P^{1/2} H_t \|_2$. Then this signal $\hat{x}_t$ is (ϵ, δ)-differentially private for the adjacency relation (6.2) with $p = 2$ if $C_2 = B_2/(1 - \rho)$, and for the adjacency relation (6.1) with $p = 2$ if $C_2 = \frac{K_2}{|\rho - \alpha|} \left(\sum_{t \geq 0} (\rho^t - \alpha^t)^2 \right)^{1/2}$.*

Proof From the bounds (6.38) or (6.39), we deduce by Theorem 2.3 that $P^{1/2} z_t + \zeta_t$ is a differentially private signal, where ζ_t is a Gaussian white noise with covariance matrix $\sigma^2 I$. Hence $P^{-1/2}(P^{1/2} z_t + \zeta_t)$ is also differentially private by resilience to post-processing, see Sect. 1.3.3. We define $\xi_t = P^{-1/2} \zeta_t$ in the Corollary.

Corollaries 6.7 and 6.8 give two differentially private mechanisms with output perturbation, provided we can design the matrices H_t to verify the assumptions of Corollary 6.6 with the (weighted) 1- or 2-norm on X. The next section discusses application examples for the privacy-preserving observer design methodology.

6.6   Application Examples

We briefly describe in this section two examples illustrating the results of this chapter, and refer the reader to Le Ny (2018) for additional details. Our first example concerns the estimation of link formation probabilities in a dynamic statistical model of social networks, the Dynamic Stochastic Blockmodel (Holland et al. 1983; Xu and Hero 2014). Consider a set of n nodes. Each node corresponds to an individual and can belong to one of N classes. Let θ_t^{ab} be the probability of forming an edge at time t between a node in class a and a node in class b, and let θ_t denote the vector of probabilities $[\theta_t^{ab}]_{1 \leq a, b \leq N}$. For example, edges could represent email exchanges or phone conversations. Edges are assumed to be formed independently of each other according to θ_t. Let $y_t^{ab} = \frac{m_t^{ab}}{n^{ab}}$ be the observed frequency of edges between classes a and b, where m_t^{ab} is the number of observed edges between classes a and b at time t, and n^{ab} is the maximum possible number of edges between these two classes. For simplicity, we assume that the quantities n^{ab} are publicly known, and we focus on the problem of estimating the parameters θ_t^{ab} by using the signals y_t^{ab}. The links formed between specific nodes constitute private information however, so directly releasing m_t^{ab} or y_t^{ab} or an estimate of θ_t based on these quantities is not allowed.

For n^{ab} is sufficiently large, Xu and Hero (2014) proposes an approximate model where y_t^{ab} is Gaussian, so that

$$y_t = \theta_t + v_t, \tag{6.41}$$

where v_t is a Gaussian noise vector with diagonal covariance matrix V_t. Next, define the logit of θ_t, denoted ψ_t, with entries $\psi_t^{ab} = \ln \frac{\theta_t^{ab}}{1-\theta_t^{ab}}$, which are well defined for $0 < \theta_t^{ab} < 1$. The dynamics of ψ_t is assumed to be linear

$$\psi_{t+1} = F\psi_t + w_t, \tag{6.42}$$

for some known matrix F, and for noise vectors w_t assumed to be iid Gaussian with known covariance matrix W (Xu and Hero 2014). The observation model (6.41) now becomes

$$y_t = g(\psi_t) + v_t, \tag{6.43}$$

where the components of g are given for all pairs a, b by

$$g^{ab}(\psi_t) = \frac{1}{(1 + e^{-\psi_t^{ab}})}.$$

An extended Kalman filter is proposed in Xu and Hero (2014) to estimate ψ, but to illustrate the ideas discussed in the previous sections, we design here a deterministic observer

$$\hat{\psi}_{t+1} = F\hat{\psi}_t + H(y_t - g(\hat{\psi}_t)) = (F\hat{\psi}_t - Hg(\hat{\psi}_t)) + Hy_t,$$

with H a constant square gain matrix. To enforce contraction as in Corollary 6.6, we should choose H so that $\|F - HG(\psi)\| \leq \rho$, where $G(\psi)$ is the Jacobian of g at ψ. Note that $G(\psi)$ is a square and diagonal matrix with entries $G_{ii}(\psi) = \frac{e^{-\psi^i}}{(1+e^{-\psi^i})^2}$, with i indexing the pairs (a, b). The only nonlinearity in the model (6.42), (6.43) comes from the observation model (6.43). To further simplify the following discussion, let us assume that F is also diagonal. In this case, the systems completely decouple into scalar systems and we can choose H to be diagonal as well. The observer for one of these scalar system takes the form

$$z_{t+1} = fz_t + h \times \left(y_t - \frac{1}{1 + e^{-z_t}} \right) = fz_t - \frac{h}{1 + e^{-z_t}} + hy_t, \tag{6.44}$$

where $h \in \mathbb{R}$ is the observer gain to set, $f \in \mathbb{R}_+$, $z_t \in \mathbb{R}$ is one component (a, b) of $\hat{\psi}_t$ and y_t now represents just the corresponding scalar component of the measurement vector as well. The contraction condition (6.37) reads, for some $0 < \rho < 1$,

$$-\rho \leq f - \frac{h\,e^{-z}}{(1 + e^{-z})^2} \leq \rho \tag{6.45}$$

$$\text{i.e.,} \quad f - \rho \leq \frac{h\,e^{-z}}{(1 + e^{-z})^2} \leq f + \rho. \tag{6.46}$$

The value of h should be set to satisfy this condition and will depend on the value of f and the desired contraction parameter ρ (Le Ny 2018). If for example $f = 1$ in the dynamics (6.44) and the adjacency relation considered is (6.1), then by Corollary 6.7, we can publish an ϵ-differentially private estimate of ψ by computing z_k using (6.44) and adding Laplace noise to it with parameter $b = K_1 h / (\epsilon(1 - \rho)(1 - \alpha))$. Small noise requires small values of h and of ρ. Since we must take $\rho < 1$, we cannot enforce the left inequality of (6.46) for all values of z. Suppose then that we want to design a privacy-preserving observer assuming that θ remains in the interval $[0.1, 0.9]$, or equivalently $\psi \in [-2.197, 2.197]$ approximately. In this interval, we have $0.09 \leq \frac{e^{-\psi}}{(1 + e^{-\psi})^2} \leq \frac{1}{4}$, and so ρ and h must also satisfy

$$\frac{f - \rho}{0.09} \leq h \leq 4(f + \rho), \quad \text{i.e.,} \quad \frac{1 - \rho}{0.09} \leq h \leq 4(1 + \rho). \tag{6.47}$$

We can then set $h = (1 - \rho)/0.09$ to satisfy the left inequality in (6.47) with equality for the value of the contraction parameter ρ that we want to achieve. For faster observer convergence we should try to achieve the lowest possible value of ρ, although this might amplify the steady-state variance due to the unmodeled measurement noise. The inequalities (6.47) can only be satisfied for $\rho \gtrsim 0.47$, a contraction parameter that can then be achieved by taking $h \approx 5.88$.

Figure 6.1 illustrates the behavior of the privacy-preserving observer, when the privacy parameters are $\epsilon = \ln(3)$, $\delta = 0$ and $K_1 = 3 \times 10^{-3}$ and $\alpha = 0.25$ in (6.1). Moreover, as we often saw in this monograph, we should generally further filter the differentially private signal produced above, since this signal is directly perturbed by the privacy-preserving noise. In this case, one can interpret the private estimate $\tilde{z}_t = z_t + \xi_t$, with ξ the Laplace noise as in Corollary 6.7, as a noisy measurement of ψ, now with a trivial linear measurement model, in contrast to (6.43). A possible simple post-filter smoothing $\tilde{z}_t$ can then be the linear observer

$$\hat{\psi}_{t+1} = f\hat{\psi}_t + k_{\text{post}}(\tilde{z}_t - f\hat{\psi}_t), $$

and Fig. 6.1 also represents $\hat{\theta}_t = g(\hat{\psi}_t)$ for the gain value $k_{\text{post}} = 0.4$.

Our second example is concerned with syndromic surveillance systems, which monitor health-related data in real-time in a population to facilitate early detection of epidemic outbreaks (Lawson and Kleinman 2005). To estimate the state of an illness in a population, we rely here on the classical SIR model of Kermack and McKendrick (1927), see also Brauer et al. (2008), which models the evolution of an epidemic by

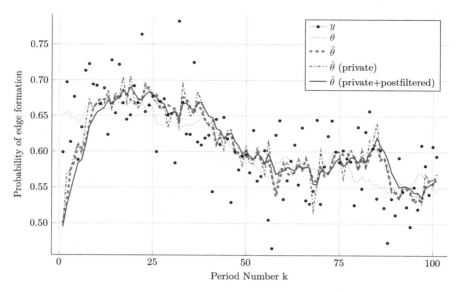

Fig. 6.1 Sample path of the estimate of the edge formation probability θ_t^{ab}, for some classes (a, b). The measured edge density is generated from one component of the model (6.41), (6.42) with $f = 1$ and w_t, v_t iid Gaussian random variables with zero mean and standard deviation 0.03 and 0.04 respectively. The trajectory θ (dotted line) starts at the value 0.65, and the observers are all initialized at the value 0.50. The upper bound ρ on the contraction rate of the observer (6.44) is set to $\rho = 0.9$, with corresponding gain $h = 1.11$. The dot-dashed line shows $1/(1 + \exp(-\bar{z}_t))$ as our private estimate of θ_t, where $\bar{z}_t$ is a ln(3)-differentially private estimate of ψ_t (hence, the estimate of θ_t is also ln(3)-differentially private), obtained by the Laplace mechanism, for the adjacency relation (6.1) with parameter values $K_1 = 3 \times 10^{-3}$, $\alpha = 0.25$. We also show a differentially private estimate obtained after further post-filtering, as explained in the text. Reproduced from Le Ny (2018) by permission of John Wiley & Sons, Ltd.

dividing individuals into 3 categories: susceptible (S), i.e., individuals who might become infected if exposed; infectious (I), i.e., currently infected individuals who can transmit the infection; and recovered (R) individuals, who are immune to the infection. A simple version of the model in continuous-time reads

$$\frac{ds}{dt} = -\mu \mathcal{R}_o i s$$
$$\frac{di}{dt} = \mu \mathcal{R}_o i s - \mu i,$$

where i and s represent the proportion of the total population in the classes I and S. The last class R is not included because we have the constraint $i + s + r = 1$. The parameter $\mathcal{R}_o$ is called the basic reproduction number and represents the average number of individuals infected by a sick person. The parameter μ represents the rate at which infectious people recover and move to the class R.

Discretizing this model with sampling period τ, we get the discrete-time model

$$s_{t+1} = s_t - \tau \mu \mathcal{R}_o i_t s_t + w_{1,t} = f_1(s_t, i_t) + w_{1,t} \tag{6.48}$$

$$i_{t+1} = i_t + \tau \mu i_t (\mathcal{R}_o s_t - 1) + w_{2,t} = f_2(s_t, i_t) + w_{2,t}, \tag{6.49}$$

where we have also introduced noise signals w_1 and w_2 in the dynamics. We assume here for simplicity that we can collect data providing a noisy measurement of the proportion of infected individuals, i.e.,

$$y_t = i_t + v_t,$$

where v_t is a noise signal. An observer can then take the form

$$\hat{s}_{t+1} = f_1(\hat{s}_t, \hat{i}_t) + h_1(y_t - \hat{i}_t)$$
$$\hat{i}_{t+1} = f_2(\hat{s}_t, \hat{i}_t) + h_2(y_t - \hat{i}_t).$$

We define the Jacobian matrix of the system (6.48), (6.49)

$$F(s, i) = I_2 + \tau \mu \mathcal{R}_o \begin{bmatrix} -i & -s \\ i & s - 1/\mathcal{R}_o \end{bmatrix},$$

as well as the gain matrix $H = [h_1, h_2]^T$ and observation matrix $C = [0, 1]$. We then design a differentially private observer with Gaussian noise using Corollary 6.8, for the adjacency relation (6.1) with $p = 2$.

Following Corollary 6.3, the contraction rate constraint (6.37) for a 2-norm on $\mathbb{R}^2$ weighted by a matrix $P \succ 0$ can be shown to be equivalent to the following family LMIs, for all $x = (s, i)$ in the regions of $[0, 1]^2$ where we want to show contraction

$$\begin{bmatrix} \rho^2 P - F_x^T P F_x + F_x^T X C + C^T X^T F_x & C^T X^T \\ X C & P \end{bmatrix} \succeq 0, \tag{6.50}$$

where we used the notation $F_x := F(s, i)$ and defined the new variable $X = PH$. If we can find P, X satisfying these inequalities, we recover the observer gain vector simply as $H = P^{-1} X$. Moreover, we argue in Le Ny (2018) that for any given value of ρ, we can try to minimize the impact of the noise by solving the following semidefinite program

$$\min_{\Sigma \succeq 0, \lambda \geq 0, P \succ 0, X} \lambda + \nu \operatorname{Tr}(\Sigma)$$

$$\text{subject to } \begin{bmatrix} \lambda & X^T \\ X & P \end{bmatrix} \succeq 0, \begin{bmatrix} \Sigma & I_2 \\ I_2 & P \end{bmatrix} \succeq 0, \text{ and (6.50)}.$$

As an example, let us assume $\mu = 0.1, \mathcal{R}_o = 2, \tau = 0.1, K_2 = 10^{-3}, \alpha = 0.25$ in (6.1), and $\epsilon = 2, \delta = 0.05$. Although not a perfectly rigorous contraction certificate, we sample the continuous set of constraints (6.50) by sampling the set $\{(s, i) | 0.01 \leq i \leq 0.25, 0.01 \leq s \leq 1 - i\}$ at the values of s, i multiple of 0.01, to obtain a finite

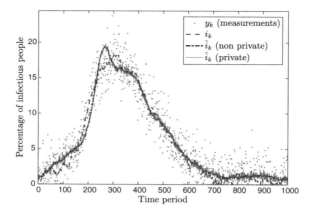

Fig. 6.2 Sample path of the estimate of the percentage of infectious people. The standard deviations for the dynamics and measurement noise were set to $\sigma_w I_2 = 0.005\sqrt{\tau} I_2$ and $\sigma_v = 0.02$ respectively. The signals were truncated to maintain positive values for i, s, y in the simulation. The true proportion of infectious people starts at 0.5%, whereas the estimate used to initialize the observer is 1%. Reproduced from Le Ny (2018) by permission of John Wiley & Sons, Ltd

number of LMIs. A more rigorous approach to enforce these constraints could make use of sum-of-squares programming (Aylward et al. 2008). Following the procedure above, for the choice $\rho = 0.996$, we obtain the observer gain $H = [3.9304; 0.2003]$ and the covariance matrix Σ with $\Sigma^{1/2} = \begin{bmatrix} 691 & 22 \\ 22 & 17 \end{bmatrix} \times 10^{-4}$ for the privacy-preserving Gaussian noise. Sample trajectories of the non-private and private (non-smoothed) estimates of i are shown on Fig. 6.2.

References

Aghannan N, Rouchon P (2003) An intrinsic observer for a class of Lagrangian systems. IEEE Trans Autom Control 48(6):936–945

Angeli D (2000) A Lyapunov approach to incremental stability properties. IEEE Trans Autom Control 47(3):410–421

Aylward EM, Parrilo PA, Slotine J-JE (2008) Stability and robustness analysis of nonlinear systems via contraction metrics and SOS programming. Automatica 44(8):2163–2170

Brauer F, van den Driessche P, Wu J (2008) Mathematical epidemiology. Lecture notes in mathematics, vol 1945. Springer, Berlin

Chung S-J et al (2008) Phase synchronization control of complex networks of Lagrangian systems on adaptive digraphs. Automatica 49(5):1148–1161

do Carmo MP, (1992) Riemannian geometry. Birkhäuser, Basel

Forni F, Sepulchre R (2014) A differential Lyapunov framework for contraction analysis. IEEE Trans Autom Control 59(3):614–628

Gauthier J-P, Kupka I (2001) Deterministic observation theory and applications. Cambridge University Press, Cambridge

Hartman P (1982) Ordinary differential equation, 2nd edn. Birkhäuser, Basel

Holland PW, Laskey KB, Leinhardt S (1983) Stochastic blockmodels: first steps. Soc Netw 5(2):109–137

Isidori A (2017) Lectures in feedback design for multivariable systems. Springer, Berlin

Kermack WO, McKendrick AG (1927) A contribution to the mathematical theory of epidemics. Proceedings of the royal society of London series A 115:700–721

Khalil HK (2002) Nonlinear systems. Prentice Hall, Upper Saddle River

Lakshmikantham V, Trigiante D (2002) Theory of difference equations: numerical methods and applications. Prentice Hall, Upper Saddle River

Lawson AB, Kleinman K (2005) Spatial and syndromic surveillance for public health. Wiley, New Jersey

Le Ny J (2015) Privacy-preserving nonlinear observer design using contraction analysis. In: Proceedings of the 54th conference on decision and control (CDC), Osaka, Japan, pp 4499–4504. https://doi.org/10.1109/CDC.2015.7402922

Le Ny J (2018) Differentially private nonlinear observer design using contraction analysis. Int J Robust Nonlinear Control. 10.1002/rnc.4392.eprint, https://onlinelibrary.wiley.com/doi/pdf/10.1002/rnc.4392

Lewis DC (1949) Metric properties of differential equations. Am J Math 71(2):294–312

Lohmiller W, Slotine J-J (1998) On contraction analysis for non-linear systems. Automatica 34(6):683–696

Pavloc A, van deVouw N, Nijmeijer H (2006) Uniform output regulation of nonlinear systems: a convergent dynamics approach. Birkhäuser, Basel

Russo G, di Bernardo M, Sontag ED (2010) Global entrainment of transcriptional systems to periodic inputs. PLOS Comput Biol 6:4

Särkkä S (2013) Bayesian filtering and smoothing. Cambridge University Press, Cambridge

Shen Z (2001) Lectures on Finsler geometry. World Scientific, Singapore

Sontag ED, (2010) Contractive systems with inputs. In: Willems J, et al (2010) Perspectives in mathematical system theory, control, and signal processing. Springer, Berlin, pp 217–228

Sontag ED, Wang Y (1997) Output-to-state stability and detectability of nonlinear systems. Syst Control Lett 29(5):279–290

Xu KS, Hero III AO (2014) Dynamic stochastic block models for time evolving social networks. J Sel Top Signal Process 8(4):552–562. Special issue on signal processing for social networks

Chapter 7
Conclusion

This monograph has introduced a number of techniques for privacy-preserving signal processing, enforcing a state-of-the-art notion of privacy, differential privacy. Let us now summarize some of the main ideas and briefly suggest some directions for future work.

The distinctive feature of the differential privacy definition, as explained in Chap. 1, is to prevent an adversary from detecting certain variations in a sensitive dataset, based on observing the output of a differentially private algorithm. The canonical example is to prevent inferring if the data of a specific individual is in a dataset or not, but as we have seen, the idea is more general and it can be useful to relax or adapt the type of variations we wish to obfuscate, which we do through the notion of adjacency relation. A significant practical advantage of differential privacy is that it avoids the need to model the auxiliary information that the adversary has access to in order to make inferences, which in most cases cannot be predicted, especially since this information can become available after the result of an analysis is published. This advantage is crucial, given that most privacy attacks are linkage attacks, i.e., are carried out by correlating information from multiple datasets.

In Chaps. 2 and 3, we discuss first how to adapt the standard Laplace and Gaussian mechanisms to publish differentially private signals by simply adding white noise. When sensitive signals need to be processed to release real-time statistics of interest, this noise can be added directly on the input signals or after processing, leading to the input and output perturbation mechanisms. Input perturbation is the easiest scheme to implement and has the advantage of releasing directly a differentially private version of the original signals. However, in many cases it can lead to an unacceptable level of degradation in the data. Output perturbation adapts the level of noise to the filtering task through the notion of sensitivity of a filter. This can lead to better performance but relies on being able to compute the sensitivity, which partly depends on the chosen definition of adjacency. Finally, by remarking that output perturbation mechanisms leave raw white noise on the output signals, which should presumably be further filtered, we were led to consider two-stage architectures, which add noise between a signal shaping filter and a post-processing filter. Again, the

J. Le Ny, *Differential Privacy for Dynamic Data*,
SpringerBriefs in Control, Automation and Robotics,
https://doi.org/10.1007/978-3-030-41039-1_7

tractability of the problem of optimizing these architectures depends on the ability to handle a sensitivity calculation, and hence on the definition of adjacency relation. These ideas were illustrated by presenting the zero-forcing equalization mechanism.

In Chaps. 4, 5 and 6 we are concerned with the problem of designing differentially private model-based estimators, using input perturbation, output perturbation or two-stage architectures. Leveraging publicly known models of the sensitive data is relatively unexplored in the literature on differential privacy, but is of course a core idea in signal processing and control. Two fundamental classes of models used for signal estimation, namely, wide-sense stationary stochastic models with known second order statistics, and linear state-space models, are discussed in Chaps. 4 and 5 respectively. For example, many privacy-sensitive datasets contain location traces, which are constrained by physical kinematic models, and it is natural to leverage such models for an estimation task, as it can significantly improve the performance/privacy tradeoffs. For the two-stage architecture, the modeling assumptions drive the choice of post-processing filter. The optimization problem also becomes more challenging, leaving some currently open problems, e.g., for further optimizing Wiener and Kalman filtering-based mechanisms. Two-stage architectures have not been systematically discussed in Chap. 6 for nonlinear model-based filters, since simply adjusting the sensitivity/performance tradeoff for output perturbation is already challenging in this case. Hence, much work still needs to be done for the design of nonlinear differentially private filters.

Finally, this monograph offers an introduction to some useful techniques for privacy-preserving signal processing, but a number of fundamental issues remain to be explored. Certain variations of the differential privacy definition could prove to be more tractable to optimize filtering mechanisms, and replacing the notion of global sensitivity by more local versions should be useful to improve performance, especially for nonlinear systems. Using other perturbation mechanisms such as subsampling and quantization in addition to noise has not been systematically explored. The filtering techniques proposed here need to be tested in practical applications. Lastly, a better understanding of the fundamental privacy-utility tradeoffs, as opposed to the tradeoffs offered by specific choices of architecture, is necessary to guide future work on this topic.

Appendix A
Discrete-Time Stochastic Signals

In this appendix we recall some basic notions from statistical signal processing. Given a discrete-time stochastic vector-valued signal $u = \{u_t\}_{t=-\infty}^{\infty}$, with $u_t \in \mathbb{R}^d$ for some integer d,[1] we define its mean function as $m_t^u = \mathbb{E}[u_t]$ and its autocorrelation function as $R_{t_1,t_2}^u = \mathbb{E}[u_{t_1} u_{t_2}^T]$. The signal u is then called wide-sense stationary (WSS) if the following conditions are satisfied:

1. The mean function is constant, i.e., $m_t^u = m^u$, $\forall t$.
2. The autocorrelation R_{t_1,t_2}^u depends only on $t_1 - t_2$, so that we can define the autocorrelation function R_t^u from $R_{t_1,t_2}^u = R_{t_1-t_2}^u$, by a slight abuse of notation. In particular, $R_\tau^u = \mathbb{E}[u_\tau u_{\tau-t}^T]$, for any τ.
3. The variance of the signal is finite, i.e., $\mathbb{E}[|u_t - m_t^u|^2] = \mathrm{Tr}(R_0^u) - |m^u|^2 < \infty$, for all t.

Let u be a WSS process with real-valued components. First, note that have $R_{-t}^u = \mathbb{E}[u_{\tau+t} u_\tau^T]^T = (R_t^u)^T$. The z-spectrum matrix of u is defined as the double-sided z-transform of the autocorrelation function and denoted $P_u(z) = \sum_{t=-\infty}^{\infty} R_t^u z^{-t}$. The power spectrum is then defined as $P_u(e^{j\omega})$, assuming the sequence defining the z-spectrum converges on the unit circle.[2] We have

$$P_u(e^{j\omega})^* = \sum_{t=-\infty}^{\infty} (R_t^u)^T e^{j\omega t} = \sum_{t=-\infty}^{\infty} R_{-t}^u e^{j\omega t} = P_u(e^{j\omega}). \tag{A.1}$$

---

[1] These definitions extend to complex valued-signals by replacing transposition by conjugate transposition (Hayes 1996; Kailath et al. 2000) and expressions of the form $F(z^{-1})^T$ by the parahermitian conjugate $[F(1/z^*)]^*$, but we focus here on real-valued signals for simplicity.

[2] This is the case for example if there exists a positive definite matrix R and some positive scalar $\alpha < 1$ such that $R_t^u \prec R\alpha^{|t|}$, for all $-\infty < t < \infty$. The Wiener–Khinchin theorem provides more general conditions on the autocorrelation sequence under which a notion of spectrum can be defined, but this will not be discussed here.

© The Author(s), under exclusive license to Springer Nature Switzerland AG 2020
J. Le Ny, *Differential Privacy for Dynamic Data*,
SpringerBriefs in Control, Automation and Robotics,
https://doi.org/10.1007/978-3-030-41039-1

More generally the z-spectrum for a signal u with real-valued components satisfies the relation $P_u(z^{-1})^T = P_u(z)$. We see from (A.1) that the power spectrum matrices are Hermitian, and in fact they are even positive semidefinite, for all ω in $[-\pi, \pi)$, see Kailath et al. (2000, Sect. 6.6).

Given two vector-valued signals u and v with real-valued components, we denote the cross-correlation matrix sequence $R_{s,t}^{uv} = \mathbb{E}[u_s v_t^T]$. Similarly to the case of a single signal, we say that u and v are *jointly* WSS if they are each WSS and their cross-correlation $R_{s,t}^{uv}$ depends only on $s - t$, in which case we define $R_t^{uv} = \mathbb{E}[u_\tau v_{\tau-t}^T]$, for any τ. We also have $R_{-t}^{uv} = \left(R_t^{vu}\right)^T$. For jointly WSS signals, the cross power spectral density matrix is defined as $P_{uv}(e^{j\omega}) = \sum_{t=-\infty}^{\infty} R_t^{uv} e^{-j\omega t}$ and the z-cross-spectrum as $P_{uv}(z) = \sum_{t=-\infty}^{\infty} R_t^{uv} z^{-t}$. Note that from the relation between cross-correlation matrices, we get $P_{vu}(z) = P_{uv}(z^{-1})^T$ and so $P_{vu}(e^{j\omega}) = P_{uv}(e^{j\omega})^*$.

An important situation where we have two jointly WSS signals u and v is if u is WSS and $v = Fu$, where F is a linear time-invariant (LTI) system with square integrable impulse response $\{F_t\}_{t\geq 0}$ and transfer matrix $F(z)$ (and zero initial conditions). In this case, we have $R_t^{uv} = R_t^u * F_{-t}^T = \sum_\alpha R_{t+\alpha}^u F_\alpha^T$ and $P_{uv}(z) = P_u(z)F(z^{-1})^T$, $P_{uv}(e^{j\omega}) = P_u(e^{j\omega})F(e^{j\omega})^*$. Moreover, we have

$$R_t^v = F_t * R_t^{uv} = F_t * R_t^u * F_{-t}^T,$$
$$P_v(z) = F(z)P_{uv}(z) = F(z)P_{uv}(z)F(z^{-1})^T,$$
$$P_v(e^{j\omega}) = F(e^{j\omega})P_{uv}(e^{j\omega}) = F(e^{j\omega})P_u(e^{j\omega})F(e^{j\omega})^*.$$

For simplicity in this book we assume when needed that the z-spectrum of our signals have components that are rational functions of z. This holds if the input signal u on Fig. 3.2 has rational spectrum and the filters are finite-dimensional LTI systems (hence, with rational transfer functions). In particular, we do not discuss the case of spectra with impulses (Papoulis 1985), which correspond to periodic components in a signal. Moreover, we also assume when needed that the power spectrum of our signals is positive, i.e., the matrix $P_u(e^{j\omega})$ is positive definite for all $\omega \in [-\pi, \pi)$.

References

Hayes MH (1996) Statistical digital signal processing and modeling. Wiley, New Jersey

Kailath T, Sayed AH, Hassibi B (2000) Linear estimation. Prentice Hall, Upper Saddle River

Papoulis A (1985) Predictable processes and wold's decomposition: a review. IEEE Trans Acoust, Speech, Signal Process 33(4):933–938

Appendix B
Proof of the Basic Composition Theorem

In this appendix we provide a proof of Theorem 1.2. We use the following terminology and notation. We fix a generic probability space $(\Omega, \mathscr{F}, \mathbb{P})$, and consider a measurable space $(\mathsf{Y}, \mathscr{Y})$. Let U be a space equipped with an adjacency relation Adj, as defined in Sect. 1.3.1. In that section, the notion of mechanism M is also introduced informally as a randomized map from U to R. More formally, a mechanism M is a map from $\mathsf{U} \times \Omega$ to Y, such that for each element u in U, $M(u, \cdot)$ is a random variable with values in Y. This random variable is also denoted $M(u)$, following the standard practice in probability of omitting the mention of the elements ω of the sample space Ω.

The proof of Theorem 1.2 requires first a lemma providing alternative characterizations of differential privacy. Recall that a signed measure on $\mathscr{Y}$ is a function from $\mathscr{Y}$ to $[-\infty, +\infty]$ that is countably additive (Dudley 2002, p. 178). We say that the signed measure ν is bounded above by δ if it satisfies $\nu(S) \leq \delta$ for all $S \in \mathscr{Y}$, and that it is nonnegative if $\nu(S) \geq 0$ for all $S \in \mathscr{Y}$. We denote by $\mathscr{M}_b^+(\mathsf{Y})$ the space of measurable bounded and nonnegative functions on Y and we define $\mu g := \int g \, d\mu$ for $g \in \mathscr{M}_b^+(\mathsf{Y})$ and μ a nonnegative measure on $\mathscr{Y}$.

Lemma B.1 *Let $\epsilon \geq 0$, $\delta \in [0, 1]$. Let U be a space equipped with an adjacency relation Adj, $(\mathsf{Y}, \mathscr{Y})$ a measurable space, and $M : \mathsf{U} \times \Omega \to \mathsf{Y}$ a mechanism. The following are equivalent:*

1. *M is (ϵ, δ)-differentially private for Adj, satisfying (1.1).*
2. *For any u, u' in U such that $Adj(u, u')$, there exists a nonnegative measure $\mu^{u,u'}$ on $(\mathsf{Y}, \mathscr{Y})$, bounded above by δ, such that we have, for all S in $\mathscr{Y}$,*

$$\mathbb{P}(M(u) \in S) \leq e^\epsilon \, \mathbb{P}(M(u') \in S) + \mu^{u,u'}(S). \tag{B.1}$$

3. *For any u, u' in U such that $Adj(u, u')$, there exists a nonnegative measure $\mu^{u,u'}$ on $(\mathsf{Y}, \mathscr{Y})$, bounded above by δ, such that for all $g \in \mathscr{M}_b^+(\mathsf{R})$, we have*

$$\mathbb{E}(g(M(u))) \leq e^\epsilon \, \mathbb{E}(g(M(u'))) + \mu^{u,u'} g. \tag{B.2}$$

© The Author(s), under exclusive license to Springer Nature Switzerland AG 2020
J. Le Ny, *Differential Privacy for Dynamic Data*,
SpringerBriefs in Control, Automation and Robotics,
https://doi.org/10.1007/978-3-030-41039-1

Proof $1 \Rightarrow 2$. Suppose that M is (ϵ, δ)-differentially private. For u, u' adjacent, define the signed measure $\nu^{u,u'}$ by $S \mapsto \nu^{u,u'}(S) := \mathbb{P}(M(u) \in S) - e^{\epsilon}\mathbb{P}(M(u') \in S)$. By the definition (1.1), $\nu^{u,u'}$ is bounded above by δ. Let $\mu^{u,u'}$ be the positive variation of $\nu^{u,u'}$, i.e., $\mu^{u,u'}(S) := \sup\{\nu^{u,u'}(G) : G \in \mathcal{Y}, G \subset S\}$, for all $S \in \mathcal{Y}$. Then $\mu^{u,u'}$ is a nonnegative measure (Dudley 2002, Theorem 5.6.1), and is bounded above by δ since $\nu^{u,u'}$ is. Since $\nu^{u,u'}(S) \leq \mu^{u,u'}(S)$ for all $S \in \mathcal{Y}$, we have (B.1).

$2. \Rightarrow 3$. Let B be an upper bound on g, i.e., $0 \leq g(r) \leq B$, for all $r \in \mathbb{R}$. For any $k \geq 1$, we divide the interval $[0, B]$ into k consecutive intervals I_i, $1 \leq i \leq k$, of length B/k, and we let $A_i = g^{-1}(I_i)$ and c_i be the mid-point of the interval I_i. Then (B.2) holds for the simple function $g_k := \sum_{i=1}^{k} c_i 1_{A_i}$, and these functions approximate g. We conclude using the dominated convergence theorem.

$3. \Rightarrow 1$. Take $g = 1_S$ and use the fact that $\mu^{u,u'}$ is bounded above by δ.   □

Recall that a probability kernel between two measurable spaces $(\mathsf{Y}_1, \mathcal{Y}_1)$ and $(\mathsf{Y}, \mathcal{Y})$ is a function $k : \mathsf{Y}_1 \times \mathcal{Y} \to [0, 1]$ such that $k(\cdot, S)$ is measurable for each $S \in \mathcal{Y}$ and $k(r, \cdot)$ is a probability measure on $(\mathsf{Y}, \mathcal{Y})$ for each $r \in \mathsf{Y}_1$. The following result is a somewhat more formal version of Theorem 1.2.

Theorem B.1 (Basic composition theorem for differential privacy) *Let $\epsilon_1, \epsilon_2 \geq 0, \delta_1, \delta_2 \in [0, 1]$. Let U be a space with an adjacency relation Adj. Let $(\mathsf{Y}_1, \mathcal{Y}_1)$ and $(\mathsf{Y}_2, \mathcal{Y}_2)$ be measurable spaces and denote $\mathcal{Y}_1 \otimes \mathcal{Y}_2$ the product σ-algebra on $\mathsf{Y}_1 \times \mathsf{Y}_2$. Let $M_1 : \mathsf{U} \times \Omega \to \mathsf{Y}_1$ be an (ϵ_1, δ_1)-differentially private mechanism for Adj. Let $M_2 : \mathsf{U} \times \Omega \to \mathsf{Y}_2$ be another mechanism, such that for each $u \in \mathsf{U}$, there exists a probability kernel $k^u : \mathsf{Y}_1 \times (\mathcal{Y}_2 \otimes \mathcal{Y}_2) \to [0, 1]$ for the conditional probabilities $\mathbb{P}((M_1(u), M_2(u)) \in S | M_1(u) = m_1) := k^u(m_1, S)$, which satisfy, for all $S \in \mathcal{Y}_1 \otimes \mathcal{Y}_2$, $m_1 \in \mathsf{Y}_1$ and u, u' adjacent in U,*

$$k^u(m_1, S) \leq e^{\epsilon_2} k^{u'}(m_1, S) + \delta_2. \tag{B.3}$$

Then $M(u) := (M_1(u), M_2(u))$ is $(\epsilon_1 + \epsilon_2, \delta_1 + \delta_2)$-differentially private.

The condition (B.3) on the conditional probabilities of $M_2(u)$ given $M_1(u)$ says that as u varies but for a given a value m_1 taken by the first mechanism M_1, the remaining variations in the distribution of M_2 should satisfy an (ϵ_2, δ_2)-differential privacy property. A special case is when the distribution of $M_2(u)$ given $M_1(u) = m_1$ does not depend on u, i.e., $k^u(m_1, S) \equiv k(m_1, S)$ for all u and all m_1, S, for some kernel k, so that we can take $\epsilon_2 = \delta_2 = 0$ in (B.3). This case corresponds to the mechanism M_2 producing its output based on $M_1(u)$ without re-accessing the original dataset u, i.e., simply post-processing the result of M_1. The composition theorem is then called the *resilience to post-processing* property of differential privacy: if publishing the output of M_1 is (ϵ, δ)-differentially private, then so is publishing any additional output computed based on the result of M_1 without directly re-accessing the sensitive dataset. Hence, post-processing can be used freely to improve the *accuracy* of an output, as done in many places in this monograph, without worrying about a possible loss of privacy. Similarly, an adversary processing a differentially private output without accessing the original data cannot weaken the guarantee.

Proof Consider two adjacent elements u, u' in U. First, note that since $k^u(m_1, S)$ takes value in $[0, 1]$, the left hand side of inequality (B.3) is upper bounded by 1 and hence without loss of generality we can replace this inequality by

$$k^u(m_1, S) \leq \min\{e^{\epsilon_2}k^{u'}(m_1, S), 1\} + \delta_2.$$

Then, by the disintegration formula, for all sets $S \in \mathscr{Y}_1 \otimes \mathscr{Y}_2$,

$$\mathbb{P}((M_1(u), M_2(u)) \in S) = \int_{\mathsf{Y}_1} k^u(m_1, S)\, \mathbb{P}(M_1(u) \in dm_1)$$

$$\leq \int_{\mathsf{Y}_1} \min\{e^{\epsilon_2}k^{u'}(m_1, S), 1\}\, \mathbb{P}(M_1(u) \in dm_1) + \delta_2.$$

Next, we bound the first term in the last inequality using the characterization 3 in Lemma B.1 of the (ϵ_1, δ_1)-differential privacy property of M_1. Denoting $g(m_1) = \min\{e^{\epsilon_2}k^{u'}(m_1, S), 1\}$, we have for some measure $\mu_1^{u,u'}$ bounded above by δ_1

$$\int_{\mathsf{Y}_1} g(m_1)\mathbb{P}(M_1(u) \in dm_1) \leq e^{\epsilon_1} \int_{\mathsf{Y}_1} g(m_1)\mathbb{P}(M_1(u') \in dm_1) + \int_{\mathsf{Y}_1} g(m_1)\mu_1^{u,u'}(dm_1)$$

$$\leq e^{\epsilon_1} \int_{\mathsf{Y}_1} g(m_1)\mathbb{P}(M_1(u') \in dm_1) + \delta_1,$$

using the fact that g is bounded above by 1. Finally, since we also have $g(m_1) \leq e^{\epsilon_2}k^{u'}(m_1, S)$, we obtain

$$\mathbb{P}((M_1(u), M_2(u)) \in S) \leq e^{\epsilon_1+\epsilon_2} \int_{\mathsf{Y}_1} k^{u'}(m_1, S)\, \mathbb{P}(M_1(u') \in dm_1) + \delta_1 + \delta_2$$

$$= e^{\epsilon_1+\epsilon_2}\, \mathbb{P}((M_1(u'), M_2(u')) \in S) + \delta_1 + \delta_2.$$

Since this is valid for all measurable sets S and all adjacent pairs u, u', the composed mechanism M is $(\epsilon_1 + \epsilon_2, \delta_1 + \delta_2)$-differentially private. $\qquad\square$

Reference

Dudley RM (2002) Real analysis and probability, 2nd edn. Cambridge University Press, Cambridge

Series Editor Biographies

Tamer Başar is with the University of Illinois at Urbana-Champaign, where he holds the academic positions of Swanlund Endowed Chair, Center for Advanced Study (CAS) Professor of Electrical and Computer Engineering, Professor at the Coordinated Science Laboratory, Professor at the Information Trust Institute, and Affiliate Professor of Mechanical Science and Engineering. He is also the Director of the Center for Advanced Study—a position he has been holding since 2014. At Illinois, he has also served as Interim Dean of Engineering (2018) and Interim Director of the Beckman Institute for Advanced Science and Technology (2008–2010). He received the B.S.E.E. degree from Robert College, Istanbul, and the M.S., M.Phil., and Ph.D. degrees from Yale University. He has published extensively in systems, control, communications, networks, optimization, learning, and dynamic games, including books on non-cooperative dynamic game theory, robust control, network security, wireless and communication networks, and stochastic networks, and has current research interests that address fundamental issues in these areas along with applications in multi-agent systems, energy systems, social networks, cyber-physical systems, and pricing in networks.

In addition to his editorial involvement with these Briefs, Başar is also the Editor of two Birkhäuser series on *Systems and Control: Foundations and Applications* and *Static and Dynamic Game Theory: Foundations and Applications*, the Managing Editor of the *Annals of the International Society of Dynamic Games* (ISDG), and member of editorial and advisory boards of several international journals in control, wireless networks, and applied mathematics. Notably, he was also the Editor-in-Chief of *Automatica* between 2004 and 2014. He has received several awards and recognitions over the years, among which are the Medal of Science of Turkey (1993); Bode Lecture Prize (2004) of IEEE CSS; Quazza Medal (2005) of IFAC; Bellman Control Heritage Award (2006) of AACC; Isaacs Award (2010) of ISDG; Control Systems Technical Field Award of IEEE (2014); and a number of international honorary doctorates and professorships. He is a member of the US National Academy of Engineering, a Life Fellow of IEEE, Fellow of IFAC, and Fellow of SIAM. He has served as an IFAC Advisor (2017-), a Council Member of IFAC (2011–2014),

J. Le Ny, *Differential Privacy for Dynamic Data*,
SpringerBriefs in Control, Automation and Robotics,
https://doi.org/10.1007/978-3-030-41039-1

president of AACC (2010–2011), president of CSS (2000), and founding president of ISDG (1990–1994).

Miroslav Krstic is Distinguished Professor of Mechanical and Aerospace Engineering, holds the Alspach endowed chair, and is the founding director of the Cymer Center for Control Systems and Dynamics at UC San Diego. He also serves as Senior Associate Vice Chancellor for Research at UCSD. As a graduate student, Krstic won the UC Santa Barbara best dissertation award and student best paper awards at CDC and ACC. Krstic has been elected Fellow of IEEE, IFAC, ASME, SIAM, AAAS, IET (UK), AIAA (Assoc. Fellow), and as a foreign member of the Serbian Academy of Sciences and Arts and of the Academy of Engineering of Serbia. He has received the SIAM Reid Prize, ASME Oldenburger Medal, Nyquist Lecture Prize, Paynter Outstanding Investigator Award, Ragazzini Education Award, IFAC Nonlinear Control Systems Award, Chestnut textbook prize, Control Systems Society Distinguished Member Award, the PECASE, NSF Career, and ONR Young Investigator awards, the Schuck ('96 and '19) and Axelby paper prizes, and the first UCSD Research Award given to an engineer. Krstic has also been awarded the Springer Visiting Professorship at UC Berkeley, the Distinguished Visiting Fellowship of the Royal Academy of Engineering, and the Invitation Fellowship of the Japan Society for the Promotion of Science. He serves as Editor-in-Chief of *Systems and Control Letters* and has been serving as Senior Editor for *Automatica* and *IEEE Transactions on Automatic Control*, as editor of two Springer book series—*Communications and Control Engineering* and *SpringerBriefs in Control, Automation and Robotics*—and has served as Vice President for Technical Activities of the IEEE Control Systems Society and as chair of the IEEE CSS Fellow Committee. Krstic has coauthored thirteen books on adaptive, nonlinear, and stochastic control, extremum seeking, control of PDE systems including turbulent flows, and control of delay systems.